THE PICKERING MASTERS

THE WORKS OF CHARLES BABBAGE

Volume 8. *The economy of machinery and manufactures*

THE WORKS OF
CHARLES BABBAGE

EDITED BY

MARTIN CAMPBELL-KELLY

VOLUME
8

THE ECONOMY OF MACHINERY AND MANUFACTURES

ALDI DISCIP ANGLVS

ROUTLEDGE

Routledge
Taylor & Francis Group

LONDON

First published 1989 by Pickering & Chatto (Publishers) Limited

Published 2016 by Routledge

2 Park Square, Milton Park, Abingdon, Oxon OX14 4RN

Routledge is an imprint of the Taylor & Francis Group, an informa business

Babbage, Charles
The works of Charles Babbage
1. Mathematics
I. Title II. Campbell-Kelly, Martin
510 QA37.2

ISBN 13: 978-1-138-76377-7 (hbk)

INTRODUCTION

On the economy of machinery and manufactures is Babbage's most important book. This classic work, whose first edition appeared in 1832, established Babbage's reputation as a major pundit and commentator on industry. The book was unique in its focus on manufacture and was a turning point in economic writing. A critical analysis of the book is given by Maxine Berg in the *General Introduction* to the *Works of Babbage*, Vol. 1.

The *Economy of machinery and manufactures* appeared in embryonic form first as an article 'On the general principles which regulate the application of machinery to manufactures and the mechanical arts' in the *Encyclopedia Metropolitana* (1829), and subsequently in four editions. There were many reprintings and translations.[a]

Although there were several editions, and – unusually for Babbage – he polished and slightly reworded his prose from time to time, he removed nothing of significance between successive editions. Thus each edition may be regarded as a proper superset of the preceding one, so that the 4th edition, reprinted here, subsumes all of Babbage's writings on the economy of machinery and manufactures up to 1835.

Because Babbage added small fragments from one edition to the next, as well as complete chapters, and reordered some material, the exact genesis of each chapter cannot be succinctly stated. The approximate evolution of the book may, however, be given as follows.

The *Economy of machinery and manufactures* consists of two

[a] First edition: 1832; German translation, Berlin, 1833. Second edition: 1832; Philadelphia, 1832. Third edition: 1833; French translation, Brussels, 1833, Paris, 1833, Brusells, 1834; Italian translation, Florence, 1834, Turin, 1863; Spanish translation, Madrid, 1835; Swedish translation, 1835; Russian translation, 1835. Fourth edition: 1835; reprinted 1841, 1846.

sections. Section I essentially describes the minutiae of manufacturing techniques. Section II, 'On the domestic and political economy of manufactures', is a much broader analysis of the economic principles underpinning manufacturing.

Section I was given the title 'On the mechanical principles' in the 1829 *Encyclopedia Metropolitana* article, but this was not used in the book. This section consists of 12 chapters, all of which appeared in the 1829 article, mostly in sections with approximately the same titles. Although material is added in the subsequent editions, this is largely additional detail rather than significant new principles.

Section II, 'On the domestic and political economy of manufactures', consists of 22 chapters. This section was much less well developed in the 1829 article, and it only provided the material for 10 chapters, 18–20, 22–23, 28–30 and 33–34.[a] Several of these chapters were extended in the first edition of 1832, which also introduced all of the remaining chapters *except* 14, 26 and 32. These last three chapters first appeared in the second edition in response to criticisms that the book had failed to address the major economic policy issues of the day. This material was also issued as a separate pamphlet, *On currency, on a new system of manufacturing, and on the effect of machinery on human labour* (1833), of which the facsimile title page is reproduced opposite. The third edition contained few alterations and the additions were not extensive. The fourth edition contained a few further alterations in the text and a minor addition described in the preface.

NOTE ON PARLIAMENTARY REPORTS

In the *Economy of machinery and manufactures*, Babbage makes extensive use of the reports of the parliamentary select committees on industrial topics and political economy. The full titles of the reports cited by Babbage, generally with informal titles, are given below. Year and volume numbers refer to *Parliamentary Papers*.

'Report from the select committee on the state of the woollen manufacture of England', 1806, Vol. 3.

'Four reports from the select committee on the expediency of

[a] Arabic numbering of chapters is used in this introduction for convenience.

ON CURRENCY,

ON

A NEW SYSTEM OF MANUFACTURING,

AND ON THE

EFFECT OF MACHINERY ON HUMAN LABOUR.

BEING THREE CHAPTERS

EXTRACTED FROM THE THIRD EDITION OF

"THE ECONOMY OF

MACHINERY AND MANUFACTURES."

BY

CHARLES BABBAGE, ESQ.

LUCASIAN PROFESSOR OF MATHEMATICS IN THE UNIVERSITY
OF CAMBRIDGE.

LONDON:
CHARLES KNIGHT, PALL MALL EAST.
1833.

confining the distilleries to the use of sugar and molasses . . .', 1808, Vol. 4.
'Two reports from the select committee on the framework knitters' petitions', 1812, Vol. 2.
'Report from the select committee on the petitions of the watchmakers of Coventry, etc.', 1817, Vol. 6.
'Report from the select committee on the duties payable on printed goods . . .', 1818, Vol. 3.
'Report from the select committee on the copyright acts', 1818, Vol. 9.
'Report from the select committee on the framework knitters' petition', 1819, Vol. 5.
'Thirteenth and fourteenth reports of the commissioners of inquiry', 1823, Vol. 7.
'Six reports from the select committee on artizans and machinery', 1824, Vol. 5.
'Report from the select committee on the laws relating to the export of tools and machinery', 1825, Vol. 5.
'Report from the select committee on the law and practice relative to granting patents for inventions', 1829, Vol. 3.
'Report from the select committee on the state of the coal trade', 1830, Vol. 8.
'Report from the select committee on the means of lessoning the evils arising from the fluctuation of employment in manufacturing districts', 1830, Vol. 10.
'Report from the select committee on the proportions of tolls which ought to be paid by carriages propelled by steam or gas . . .', 1831, Vol. 8.

The economy of machinery and manufactures

On the ECONOMY OF MACHINERY AND MANUFACTURES

By

CHARLES BABBAGE, ESQ.RE A.M.

Lucasian Professor of Mathematics in the University of Cambridge, and Member of several Academies.

FOURTH EDITION ENLARGED.

LONDON:
CHARLES KNIGHT, PALL MALL EAST.
1835.

[iii/iv]

PREFACE

The present volume may be considered as one of the consequences that have resulted from the calculating engine, the construction of which I have been so long superintending. Having been induced, during the last ten years, to visit a considerable number of workshops and factories, both in England and on the Continent, for the purpose of endeavouring to make myself acquainted with the various resources of mechanical art, I was insensibly led to apply to them those principles of generalization to which my other pursuits had naturally given rise. The increased number of curious processes and interesting facts which thus came under my attention, as well as of the reflections which they suggested, induced me to believe that the publication of some of them might be of use to persons who propose to bestow their attention on those enquiries which I have only incidentally considered. With this view it was my intention to have delivered the present work in the form of a course of lectures at Cambridge; an intention which I was subsequently induced to alter. The substance of a considerable portion of it has, however, appeared among the preliminary chapters of the mechanical part of the Encyclopedia Metropolitana.

I have not attempted to offer a complete enumeration of all the mechanical principles which regulate the application of machinery to arts and manufactures, but I have endeavoured to present to the reader those which struck me as the most important, either for understanding the actions of machines, or / for enabling the memory to classify and arrange the facts connected with their employment. Still less have I attempted to examine all the difficult questions of *political economy* which are intimately connected with such enquiries. It was impossible not to trace or to imagine, among the wide variety of facts presented to me, some principles which seemed to pervade many establishments; and having formed such conjectures, the desire to refute or to verify them, gave an additional interest to the pursuit. Several of the principles which I have proposed, appear to me to have been unnoticed before. This was particularly the case with respect to the explanation I have given of the *division of labour*; but further enquiry satisfied me that I had been anticipated by M. Gioja, and it is probable that additional research would enable me to trace most of the other principles, which I had thought original, to previous writers, to whose merit I may perhaps be unjust, from my want of acquaintance with the historical branch of the subject.

The truth however of the principles I have stated, is of much more importance than their origin; and the utility of an enquiry into them, and of establishing others more correct, if these should be erroneous, can scarcely admit of a doubt.

The difficulty of understanding the processes of manufactures has unfortunately been greatly overrated. To examine them with the eye of a manufacturer, so as to be able to direct others to repeat them, does undoubtedly require much skill and previous acquaintance with the subject; but merely to apprehend their general principles and mutual relations, is within the power of almost every person possessing a tolerable education.

Those who possess rank in a manufacturing country, can scarcely be excused if they are entirely ignorant of principles, whose development has produced its greatness. The possessors of wealth can scarcely be indifferent to processes which, nearly / or remotely have been the fertile source of their possessions. Those who enjoy leisure can scarcely find a more interesting and instructive pursuit than the examination of the workshops of their own country, which contain within them a rich mine of knowledge, too generally neglected by the wealthier classes.

It has been my endeavour, as much as possible, to avoid all technical terms, and to describe, in concise language, the arts I have had occasion to discuss. In touching on the more abstract principles of political economy, after shortly stating the reasons on which they are founded, I have endeavoured to support them by facts and anecdotes; so that whilst young persons might be amused and instructed by the illustrations, those of more advanced judgement may find subject for meditation in the general conclusions to which they point. I was anxious to support the principles which I have advocated by the observations of others, and in this respect I found myself peculiarly fortunate. The reports of committees of the House of Commons, upon various branches of commerce and manufactures, and the evidence which they have at different periods published on those subjects, teem with information of the most important kind, rendered doubly valuable by the circumstances under which it has been collected. From these sources I have freely taken, and I have derived some additional confidence from the support they have afforded to my views.*

Dorset Street
Manchester Square
8 June, 1832 /

CHARLES BABBAGE

* I am happy to avail myself of this occasion of expressing my obligations to the Right Hon. Manners Sutton, the Speaker of the House of Commons, to whom I am indebted for copies of a considerable collection of those reports.

[vi]

PREFACE

TO THE SECOND EDITION

In two months from the publication of the first edition of this volume, three thousand copies were in the hands of the public. Very little was spent in advertisements; the booksellers, instead of aiding, impeded its sale;* it formed no part of any popular series – and yet the public, in a few weeks, purchased the whole edition. Some small part of this success, perhaps, was due to the popular exposition of those curious processes which are carried on in our workshops, and to the endeavour to take a short view of the general principles which direct the manufactories of the country. But the chief reason was the commanding attraction of the subject, and the increasing desire to become acquainted with the pursuits and interests of that portion of the people which has recently acquired so large an accession of political influence.

A greater degree of attention than I had expected has been excited by what I have stated in the first edition, respecting the 'Book-trade'. Until I had commenced the chapter, 'On the separate cost of each process of a manufacture', I had no intention of alluding to that subject: but the reader will perceive that I have throughout this volume, wherever I could, employed as illustrations, objects of easy access to the reader; and, in accordance with that principle, I selected the volume itself. When I arrived at the chapter, 'On combinations of masters against the public', I was induced, for the same reason, to expose a combination connected with literature, which, in my

* I had good evidence of this fact from various quarters; and being desirous of verifying it, I myself applied for a copy at the shop of a bookseller of respectability, who is probably not aware that he refused to procure one even for its author.

opinion, is both morally and politically wrong. I entered upon this enquiry without the slightest feeling of hostility to that trade, nor have I any wish unfavourable to / it; but I think a complete reform in its system would add to its usefulness and respectability. As the subject of that chapter has been much discussed, I have thought it right to take a view of the various arguments which have been advanced, and to offer my own opinion respecting their validity – and there I should have left the subject, content to allow my general character to plead for me against insinuations respecting my motives – but as the remarks of some of my critics affect the character of another person, I think it but just to state circumstances which will clearly disprove them.

Mr Fellowes, of Ludgate Street, who had previously been the publisher of some other volumes for me, had undertaken the publication of the first edition of the present work. A short time previous to its completion, I thought it right to call his attention to the chapter in which the book-trade is discussed; with the view both of making him acquainted with what I had stated, and also of availing myself of his knowledge in correcting any accidential error as to the facts. Mr Fellowes, 'differing from me *entirely* respecting the conclusions I had arrived at', then declined the publication of the volume. If I had then chosen to apply to some of those other booksellers, whose names appear in the Committee of 'The Trade', it is probable that they also would have declined the office of publishing for me; and, had my object been to make a case against the trade, such a course would have assisted me. But I had no such feeling; and having procured a complete copy of the *whole* work, I called with it on Mr Knight, of Pall Mall East, *whom until that day I had never seen, and with whom I had never previously had the slightest communication.* I left the book in Mr Knight's hands, with a request that, when he had read it, I might be informed whether he would undertake the publication of it; and this he consented to do. Mr Knight, therefore, is so far from being responsible for a single opinion in the present volume, that he saw it only, for a short time, a few days previous to its publication. /

It has been objected to me, that I have exposed too freely the *secrets of trade*. The only *real secrets of trade* are industry, integrity, and knowledge: to the possessors of these no exposure can be injurious; and they never fail to produce respect and wealth.

The alterations in the present edition are so frequent, that I found it impossible to comprise them in a supplement. But the three new chapters, 'On money as a medium of exchange'; 'On a new system of manufacturing'; and 'On the effect of machinery in reducing the demand for labour'; will shortly be printed separately, for the use of the purchasers of the first edition.

I am inclined to attach some importance to the new system of manufacturing; and venture to throw it out with the hope of its receiving a full discussion among those who are most interested in the subject. I believe that some such system of conducting manufactories would greatly increase the productive powers of any country adopting it; and that our own possesses much greater facilities for its application than other countries, in the greater intelligence and superior education of the working classes. The system would naturally commence in some large town, by the union of some of the most prudent and active workmen; and their example, if successful, would be followed by others. The small capitalist would next join them, and such factories would go on increasing until competition compelled the large capitalist to adopt the same system; and, ultimately, the whole faculties of *every* man engaged in manufacture would be concentrated upon one object – the art of producing a good article at the lowest possible cost – whilst the moral effect on that class of the population would be useful in the highest degree, since it would render character of far greater value to the workman than it is at present.

To one criticism which has been made, this volume is perfectly open. I have dismissed the important subject of the patent-laws in a few lines. The subject presents, in my opinion, / great difficulties, and I have been unwilling to write upon it, because I do not see my way. I will only here advert to one difficulty. What constitutes an invention? Few simple mechanical contrivances are new; and most combinations may be viewed as species, and classed under genera of more or less generality; and may, in consequence, be pronounced old or new, according to the mechanical knowledge of the person who gives his opinion.

Some of my critics have amused their readers with the wildness of the schemes I have occasionally thrown out; and I myself have sometimes smiled along with them. Perhaps it were wiser for present reputation to offer nothing but profoundly meditated plans, but I do not think knowledge will be most advanced by that course; such sparks may kindle the energies of other minds more favourably circumstanced for pursuing the enquiries. Thus I have now ventured to give some speculations on the mode of blowing furnaces for smelting iron; and even supposing them to be visionary, it is of some importance thus to call the attention of a large population, engaged in one of our most extensive manufactures, to the singular fact, that four-fifths of the steam power used to blow their furnaces actually cools them.

I have collected, with some pains, the criticisms* on the first edition of this work, and have availed myself of much information which has been communicated to me by my friends, for the improvement of the present volume. If I have succeeded in expressing that I had to explain with perspicuity, I am aware that much of this clearness is due to my friend, Dr Fitton, to whom both the present and the former edition are indebted for such an examination and correction, as an author himself has very rarely the power to bestow.

22 November, 1832 /

* Several of these have probably escaped me, and I shall feel indebted to any one who will inform my publisher of any future remarks.

REPLY TO MR BABBAGE[a]

Mr Babbage having, in the preface to the second edition of his work on the *Economy of manufactures*, stated that the booksellers, 'instead of aiding, had impeded the sale', it is considered a refutation is demanded, and the following list of the number of copies of the first edition purchased by a few of the trade on speculation will prove that the assertion is entirely void of truth.

Messrs Simpkin & Co.	460	Messrs J. and A. Arch	12
Messrs Longman & Co.	450	Messrs Parbury & Co.	12
Messrs Sherwood & Co.	350	Mr Groombridge	25
Messrs Hamilton & Co.	50	Messrs Rivingtons	12
Mr James Duncan	125	Mr W. Mason	50
Messrs Whittaker & Co.	300	Mr B. Fellowes	25
Messrs Baldwin & Co.	75		
Mr Effingham Wilson	6	Total	1977
Mr J. M. Richardson	25		

The number of copies of the *second* edition of Mr Babbage's work purchased, on the day of publication, by the same booksellers, will further prove that they are not actuated by the unworthy motives he imputes to them, and forms their justification against a charge equally unjust and unmerited.

Messrs Simpkin & Co.	500	Mr Effingham Wilson	25
Messrs Longman & Co.	250	Messrs J. & A. Arch	25
Messrs Sherwood & Co.	100	Mr Groombridge	25
Messrs Hamilton & Co	50	Messrs Rivingtons	25
Mr James Duncan	100	Mr W. Mason	50
Messrs Whittaker & Co.	200	Total	1375
Messrs Baldwin & Co	25		

[a] This *Reply to Mr Babbage*, which was not part of the original work, was inserted by some booksellers into the second edition.

PREFACE

TO THE THIRD EDITION

The alterations in this third edition are few, and the additions are not extensive. The only subject upon which it may be necessary to offer any remark, is one which has already, perhaps, occupied a larger space than it deserved.

Shortly after the publication of the second edition, I received an anonymous letter, containing a printed page, entitled 'Reply to Mr Babbage'; and I was soon informed that many of the most respectable houses in the book trade inserted this paper in every copy of my work which they sold.

In the first edition, I had censured, as I think deservedly, a combination among the larger booksellers, to keep the price of books above the level to which competition would naturally reduce it; and I pointed out the evil and oppression it produced. Of the numerous critics who noticed the subject, scarcely one has attempted to defend the monopoly; and those who deny the truth of my conclusions, have not impeached the accuracy of a single figure in the statements on which they rest. I have extracted from that reply the following: /

List of the number of copies of the first edition, purchased by a few of the trade on speculation

As stated by the booksellers		*Number really purchased by the same on speculation*
Messrs Simpkin and Co.	460	100
Messrs Longman and Co.	450	50
Messrs Sherwood and Co.	350	50
Messrs Hamilton and Co.	50	8

As stated by the booksellers		*Number really purchased by the same on speculation*
Mr James Duncan	125	25
Messrs Whittaker and Co.	300	50
Messrs Baldwin and Co.	75	25
*Mr Effingham Wilson	6	6
*Mr J. M. Richardson	25	25
Messrs J. and A. Arch	12	6
*Messrs Parbury and Co.	12	12
Mr Groombridge	25	6
*Messrs Rivington	12	12
Mr W. Mason	50	25
*Mr B. Fellowes	25	25
Total	1977	425

The author of the 'Reply', although he has not actually stated that these 1977 copies were *subscribed*, has yet left the public to make that inference, and has actually suggested it by stating that this number of copies was *purchased on speculation*, a statement which would have been perfectly true if the whole number had been purchased in the first instance, and at once. On reading the paper, therefore, I wrote to my publisher to obtain a copy of the *subscription* list, from which the column annexed to the above extract has been taken.

After the day of publication, the demand for the 'Economy of Manufactures' was rapid and regular, until the whole edition was exhausted. There can, therefore, be no pretence for asserting that any copies, taken afterwards, were purchased / *on speculation* – they were purchased because the public demanded them. The two first houses on the list, for instance, *subscribed* 150, and if they will publish the dates of their orders, the world will be able to judge whether they took the remaining 760 *on speculation*.

I have put an asterisk aginst the names of five houses, whose numbers taken *on speculation* are correctly stated.

It is right that I should add, that the delay which many have experienced in procuring the volume, has arisen from the unexpected rapidity of the sale of both editions. I have made such arrangements that no disappointment of this nature is likely to arise again.

The main question, and the only important one to the public, is the combination, and the booksellers have yet advanced nothing in its defence. The principles of 'free trade', and the importance of diffusing information at a cheap rate, are now too well understood to render the result of that combination doubtful; and the wisest course in this, as in all such cases, is – timely concession to public opinion. I shall now dismiss the subject, without fear that my motives for calling attention to it can be misunderstood, and hoping that the facts which I have elicited will advance the interests of knowledge.

Dorset Street
Manchester Square
11 February, 1833 /

PREFACE

TO THE FOURTH EDITION

The present edition, besides a few alterations in the text, contains some notes for which I am indebted to the German translation of Dr Friedenberg, published at Berlin, in 1833. Whilst I cannot but feel highly gratified at the numerous translations which have been made of the 'Economy of Manufactures', I may be permitted to regret that so few of them have contained notes, criticisms, or additions to the work itself. In almost every country, peculiar processes are practised, which would illustrate the mechanical part of this volume; and the confirmation or refutation of those economical principles which I have advanced as regulating commerce and manufactures, is of importance to the happiness of all.

The translation to which I have referred is an exception, and is rendered more valuable by the enlightened views, developed in the preface, of Mr Kloden, and by the additions of some of which I have availed myself in the present edition. I have added an extensive index, which will, I hope, render the work more valuable, as one of reference.

14 January, 1835 /

CONTENTS[a]

SECTION I

INTRODUCTION

CHAPTER I

Sources of the advantages arising from machinery and manufactures

[a] In the Contents, the numbering refers to sections, not pages. Several errors in the numbering of sections have been silently corrected, and missing section-descriptions have been interpolated in angle brackets.

CHAPTER II

Accumulating power

CHAPTER III

Regulating power

CHAPTER IV

Increase and diminution of velocity

CHAPTER V

Extending the time of action of forces

CHAPTER VI

Saving time in natural operations

CHAPTER VII

Exerting forces too great for human power, and executing operations too delicate for human touch

CHAPTER VIII

Registering operations

CHAPTER IX

Economy of materials employed

CHAPTER X

Of the identity of the work when it is of the same kind, and of its accuracy when of different kinds

CHAPTER XI

Of copying

CHAPTER XII

On the method of observing manufactories

SECTION II

ON THE DOMESTIC AND POLITICAL ECONOMY OF MANUFACTURES

CHAPTER XIII

On the difference between making and manufacturing

CHAPTER XIV

Of money as a medium of exchange

CHAPTER XV

On the influence of verification on price

CHAPTER XVI

On the influence of durability on price

CHAPTER XVII

On price, as measured by money

CHAPTER XVIII

Of raw materials

CHAPTER XIX

On the division of labour

CHAPTER XXIII

On the position of large factories

CHAPTER XXIV

On over-manufacturing

CHAPTER XXV

Enquiries previous to commencing any manufactory

CHAPTER XXVI

On a new system of manufacturing

CHAPTER XXVII

On contriving machinery

CHAPTER XXVIII

Proper circumstances for the application of machinery

CHAPTER XXXIII

On the effect of taxes and of legal restrictions upon manufactures

CHAPTER XXXIV

On the exportation of machinery

CHAPTER XXXV

On the future prospects of manufactures, as connected with science

SECTION I[a]

[a] This section is given the title 'On the mechanical principles' in Babbage's 1829 *Encyclopedia Metropolitana* article.

INTRODUCTION

The object of the present volume is to point out the effects and the advantages which arise from the use of tools and machines; to endeavour to classify their modes of action; and to trace both the causes and the consequences of applying machinery to supersede the skill and power of the human arm.

A view of the mechanical part of the subject will, in the first instance, occupy our attention, and to this the first section of the work will be devoted. The first chapter of the section will contain some remarks on the general sources from whence the advantages of machinery are derived, and the succeeding nine chapters will contain a detailed examination of principles of a less general character. The eleventh chapter contains numerous subdivisions, and is important from the extensive classification it affords of the arts in which copying is so largely / employed. The twelfth chapter, which completes the first section, contains a few suggestions for the assistance of those who propose visiting manufactories.

The second section, after an introductory chapter on the difference between *making* and *manufacturing*, will contain, in the succeeding chapters, a discussion of many of the questions which relate to the political economy of the subject. It was found that the domestic arrangement, or interior economy of factories, was so interwoven with the more general questions, that it was deemed unadvisable to separate the two subjects. The concluding chapter of this section, and of the work itself, relates to the future prospects of manufactures, as arising from the application of science. /

[3/4]

CHAPTER I

SOURCES OF THE ADVANTAGES ARISING FROM MACHINERY AND MANUFACTURES

1. There exists, perhaps, no single circumstance which distinguishes our country more remarkably from all others, than the vast extent and perfection to which we have carried the contrivance of tools and machines for forming those conveniences of which so large a quantity is consumed by almost every class of the community. The amount of patient thought, of repeated experiment, of happy exertion of genius, by which our manufactures have been created and carried to their present excellence, is scarcely to be imagined. If we look around the rooms we inhabit, or through those storehouses of every convenience, of every luxury that man can desire, which deck the crowded streets of our larger cities, we shall find in the history of each article, of every fabric, a series of failures which have gradually led the way to excellence; and we shall notice, in the art of making even the most insignificant of them, processes calculated to excite our admiration by their simplicity, or to rivet our attention by their unlooked-for results.

2. The accumulation of skill and science which has been directed to diminish the difficulty of producing / manufactured goods, has not been beneficial to that country alone in which it is concentrated; distant kingdoms have participated in its advantages. The luxurious natives of the East,* and the ruder inhabitants of the African desert are alike indebted to our looms. The produce of our factories has preceded even our most enterprising travellers.† The cotton of India

* 'The Bandana handkerchiefs manufactured at Glasgow have long superseded the genuine ones, and are now consumed in large quantities both by the natives and Chinese,' Crawfurd's *Indian Archipelago*, vol. iii, p. 505. <John Crawfurd, *History of the Indian Archipelago* (3 Vols, Edinburgh, 1820).>

† Captain Clapperton, when on a visit at the court of the Sultan Bello, states, that 'provisions were regularly sent me from the sultan's table on pewter dishes with the London stamp; and I even had a piece of meat served up on a white wash-hand basin of English manufacture', Clapperton's *Journey*, p. 88.

is conveyed by British ships round half our planet, to be woven by British skill in the factories of Lancashire: it is again set in motion by British capital; and, transported to the very plains whereon it grew, is repurchased by the lords of the soil which gave it birth, at a cheaper price than that at which their coarser machinery enables them to manufacture it themselves.*

3. The large proportion of the population of this country, who are engaged in manufactures, appears from the following table deduced from a statement in an Essay on the Distribution of Wealth, by the Rev. R. Jones:[a] /

For every hundred persons employed in agriculture, there are:

	Agriculturists	*Non-agriculturists*
In Bengal	100	25
In Italy	100	31
In France	100	50
In England	100	200

The fact that the proportion of non-agricultural to agricultural persons is continually increasing, appears both from the Report of the Committee of the House of Commons upon Manufacturers' Employment, July, 1830, and from the still later evidence of the last census; from which document the annexed table of the increase of population in our great manufacturing towns, has been deduced.

Increase of population per cent

Names of places	*1801–11*	*1811–21*	*1821–31*	*Total*
Manchester	22	40	47	151
Glasgow	30	46	38	161
Liverpool†	26	31	44	138
Nottingham	19	18	25	75
Birmingham	16	24	33	90
Great Britain	14·2	15·7	15·5	52·5

* At Calicut, in the East Indies (whence the cotton cloth called calico derives its name), the price of labour is *one-seventh* of that in England, yet the market is supplied from British looms.

† Liverpool, though not itself a manufacturing town, has been placed in this list, from its connection with Manchester, of which it is the port.

[a] Richard Jones, *An essay on the distribution of wealth, and on the sources of taxation* (London, 1831).

Thus, in three periods of ten years, during each of which the general population of the country has / increased about 15 per cent, or about 52 per cent upon the whole period of thirty years, the population of these towns has, on the average, increased 132 per cent. After this statement, there requires no further argument to demonstrate the vast importance to the well-being of this country, of making the interests of its manufacturers well understood and attended to.

4. The advantages which are derived from machinery and manufactures seem to arise principally from three sources: *The addition which they make to human power. The economy they produce of human time. The conversion of substances apparently common and worthless into valuable products.*

5. *Of additions to human power.* With respect to the first of these causes, the forces derived from wind, from water, and from steam, present themselves to the mind of every one; these are, in fact, additions to human power, and will be considered in a future page: there are, however, other sources of its increase, by which the animal force of the individual is itself made to act with far greater than its unassisted power; and to these we shall at present confine our observations.

The construction of palaces, of temples, and of tombs, seems to have occupied the earliest attention of nations just entering on the career of civilization; and the enormous blocks of stone moved from their native repositories to minister to the grandeur or piety of the builders, have remained to excite the astonishment of their posterity, long after the purposes of many of these records, as well as the names of their founders, have been forgotten. The different / degrees of force necessary to move these ponderous masses, will have varied according to the mechanical knowledge of the people employed in their transport; and that the extent of power required for this purpose is widely different under different circumstances, will appear from the following experiment, which is related by M. Rondelet, *Sur L'Art de Bâtir*.[a] A block of squared stone was taken for the subject of experiment:

	lbs
1. Weight of stone	1080
2. In order to drag this stone along the floor of the quarry, roughly chiselled, it required a force equal to	758

[a] Jean-Baptiste, *Traité de l'art de bâtir* . . . (Paris, 1829).

	lbs
3. The same stone dragged over a floor of planks required	652
4. The same stone placed on a platform of wood, and dragged over a floor of planks, required	606
5. After soaping the two surfaces of wood which slid over each other, it required	182
6. The same stone was now placed upon rollers of three inches diameter, when it required to put it in motion along the floor of the quarry	34
7. To drag it by these rollers over a wooden floor	28
8. When the stone was mounted on a wooden platform, and the same rollers placed between that and a plank floor, it required	22

From this experiment it results, that the force necessary to move a stone along

	Part of its weight
The roughly chiselled floor of its quarry is nearly	2/3
Along a wooden floor	3/5 /
By wood upon wood	5/9
If the wooden surfaces are soaped	1/6
With rollers on the floor of the quarry	1/32
On rollers on wood	1/40
On rollers between wood	1/50

At each increase of knowledge, as well as on the contrivance of every new tool, human labour becomes abridged. The man who contrived rollers, invented a tool by which his power was quintupled. The workman who first suggested the employment of soap or grease, was immediately enabled to move, without exerting a greater effort, more than three times the weight he could before.*

6. *The economy of human time* is the next advantage of machinery in manufactures. So extensive and important is this effect, that we

* So sensible are the effects of grease in diminishing friction, that the drivers of sledges in Amsterdam, on which heavy goods are transported, carry in their hand a rope soaked in tallow, which they thrown down from time to time before the sledge, in order that, by passing over the rope, it may become greased.

might, if we were inclined to generalize, embrace almost all the advantages under this single head: but the elucidation of principles of less extent will contribute more readily to a knowledge of the subject; and, as numerous examples will be presented to the reader in the ensuing pages, we shall restrict our illustrations upon this point.

As an example of the economy of time, the use of gunpowder in blasting rocks may be noticed. Several pounds of powder may be purchased for a sum / acquired by a few days' labour: yet when this is employed for the purpose alluded to, effects are frequently produced which could not, even with the best tools, be accomplished by other means in less than many months.

The dimensions of one of the blocks of limestone extracted from the quarries worked for the formation of the breakwater at Plymouth, were 26½ ft long, 13 ft wide, and 16 ft deep. This mass, containing above 4,800 cubic feet, and weighing about 400 tons, was blasted three times. Two charges of 50 lbs each were successively exploded in a hole 13 feet deep, the bore being 3 inches at top and 2½ inches at bottom: 100 lbs of powder were afterwards exploded in the rent formed by those operations. Each pound of gunpowder separated from the rock two tons of matter, or nearly 4,500 times its own weight. The expense of the powder was £6, or nearly 7½*d*. per lb: the boring occupied two men during a day and a half, and cost about 9*s*.; and the value of the produce was, at that time, about £45.

7. The simple contrivance of tin tubes for speaking through, communicating between different apartments, by which the directions of the superintendent are instantly conveyed to the remotest parts of an establishment, produces a considerable economy of time. It is employed in the shops and manufactories in London, and might with advantage be used in domestic establishments, particularly in large houses, in conveying orders from the nursery to the kitchen, or from the house to the stable. Its convenience arises not merely from saving the servant or workman useless journeys to receive directions, but from / relieving the master himself from that indisposition to give trouble, which frequently induces him to forego a trifling want, when he knows that his attendant must mount several flights of stairs to ascertain his wishes, and, after descending, must mount again to supply them. The distance to which such a mode of communication can be extended, does not appear to have been ascertained, and would be an interesting subject for enquiry. Admitting it to be possible between London and Liverpool, about seventeen minutes

would elapse before the words spoken at one end would reach the other extremity of the pipe.

8. The art of using the diamond for cutting glass has undergone, within a few years, a very important improvement. A glazier's apprentice, when using a diamond set in a conical ferrule, as was always the practice about twenty years since, found great difficulty in acquiring the art of using it with certainty; and, at the end of a seven years' apprenticeship, many were found but indifferently skilled in its employment. This arose from the difficulty of finding the precise angle at which the diamond cuts, and of guiding it along the glass at the proper inclination when that angle is found. Almost the whole of the time consumed and of the glass destroyed in acquiring the art of cutting glass, may now be saved by the use of an improved tool. The gem is set in a small piece of squared brass with its edges nearly parallel to one side of the square. A person skilled in its use now files away the brass on one side until, by trial, he finds that the diamond will make a clean cut, when guided by keeping this edge pressed against a ruler. The diamond and its mounting are now / attached to a stick like a pencil, by means of a swivel allowing a small angular motion. Thus, even the beginner at once applies the cutting edge at the proper angle, by pressing the side of the brass against a ruler; and even though the part he holds in his hand should deviate a little from the required angle, it communicates no irregularity to the position of the diamond, which rarely fails to do its office when thus employed.

The relative hardness of the diamond, in different directions, is a singular fact. An experienced workman, on whose judgement I can rely, informed me that he has seen a diamond ground with diamond powder on a cast-iron mill for three hours without its being at all worn, but that, on changing its direction with respect to the grinding surface, the same edge was ground away.

9. *Employment of materials of little value.* The skins used by the goldbeater are produced from the offal of animals. The hoofs of horses and cattle, and other horny refuse, are employed in the production of the prussiate of potash, that beautiful, yellow, crystallized salt, which is exhibited in the shops of some of our chemists. The worn-out saucepans and tinware of our kitchens, when beyond the reach of the tinker's art, are not utterly worthless. We sometimes meet carts loaded with old tin kettles and worn-out iron coal-skuttles traversing our streets. These have not yet completed their useful course; the less corroded parts are cut into strips,

punched with small holes, and varnished with a coarse black varnish for the use of the trunk-maker, who protects the edges and angles of his boxes with them; the remainder are conveyed to the manufacturing chemists / in the outskirts of the town, who employ them in combination with pyroligneous acid, in making a black die for the use of calico printers.[a]

---

[a] The following appears in the section of additions, pp. [393–6], in the original: The establishments for slaughtering horses at Montfaucon, near Paris, supply another illustration of the profitable conversion of substances apparently of little value.

1. The hair is first cut off from the mane and tail. It amounts usually to about a quarter of a pound, which, at 5*d*. per lb is worth 1¼*d*.

2. The skin is then taken off, and sold fresh to the tanners of the neighbourhood. It usually weighs about 60 lb, and produces from 9*s*. to 12*s*.

3. The blood may be used either as manure, or by sugar refiners, or as food for animals. Pigeons, poultry, and especially turkeys, soon fatten upon it; but a few days before they are killed they should be fed with grain, in order to remove the disagreeable taste communicated to the flesh by the previous food. In order to use the blood for manure, it is previously / dried. The cwt costs at Paris about 9*s*., and its package and transport to America, where it is chiefly used, cost nearly as much more. A horse produces about 20 lb of dried blood, worth about 1*s*. 9*d*.

4. The shoes are removed from the dead horses, and sold, either to be used again, or as old iron. The nails also are collected and sold. The average produce of the shoes and shoe-nails of a horse is about 2½*d*.

5. The hoofs are sold partly to turners and comb-makers, partly to manufacturers of sal ammonia and Prussian blue, who pay for them about 1*s*. 5*d*.

6. The fat is very carefully collected and melted down. In lamps it gives more heat than oil, and is therefore demanded by enamellers and glass toy-makers. It is also used for greasing harness, shoe leather, etc.; for soap and for making gas: it is worth about 6*d*. per lb. A horse on an average yields 8 lb of fat, worth about 4*s*.; but well-fed horses sometimes produce nearly 60 lb.

7. The best pieces of the flesh are eaten by the workmen; the rest is employed as food for cats, dogs, pigs, and poultry. It is likewise used as manure, and in the manufacture of Prussian blue. A horse has from 300 to 400 lb of flesh, which sells for from £1 8*s*. to £1 17*s*.

8. The tendons are separated from the muscles: the smaller are sold fresh, to the glue makers in the neighbourhood; the larger are dried, and sent off in greater quantities for the same purpose. A horse yields about 1 lb of dried tendons, worth about 3*d*.

9. The bones are sold to cutlers, fan makers, and manufacturers of sal ammoniac and ivory black. A horse yields about 90 lb, which sell for 2*s*.

10. The smaller intestines are wrought into coarse strings for lathes; the larger are sold as manure.

11. Even the maggots, which are produced in great numbers in the refuse, are not lost. Small pieces of the horse flesh are piled up, about half a foot high; and being covered slightly with straw to protect them from the sun, soon allure the flies, which deposit their eggs in them. In a few days the putrid flesh is converted into a living mass of maggots. These are sold by measure; some are used for bait in fishing, but the greater part as food for fowls, and especially for pheasants. One horse yields maggots which sell for about 1*s*. 5*d*.

12. The rats which frequent these establishments are innumerable, and they have

10. *Of tools*. The difference between a *tool* and a *machine* is not capable of very precise distinction; nor is it necessary, in a popular explanation of those terms, to limit very strictly their acceptation. A *tool* is usually more *simple* than a machine; it is generally used with the hand, whilst a machine is frequently moved by animal or steam power. The simpler *machines* are often merely one or more *tools* placed in a frame, and acted on by a moving power. In pointing out the advantages of *tools*, we shall commence with some of the simplest.

11. To arrange twenty thousand needles thrown promiscuously into a box, mixed and entangled in every possible direction, in such a form that they shall be all parallel to each other, would, at first sight, appear a most tedious occupation; in fact, if each needle were to be separated individually, many hours must be consumed in the process. Yet this is an operation which must be performed many times in the manufacture of needles; and it is accomplished in a few minutes by a very simple *tool*; nothing more being requisite than a small flat tray of sheet iron, slightly concave at the bottom. In this the needles are placed, and shaken in a peculiar manner, by throwing them up a very little, and giving at the same time a slight longitudinal motion to the tray. The shape of the needles assists their arrangement; for if two

---

been turned to profit by the proprietors. The fresh carcass of a horse is placed at night in a room, which has a number of openings near the floor. The rats are attracted into it, and the openings then closed. 16,000 rats were killed in one room in four weeks, without any perceptible diminution of their number. The furriers purchase the rat skins at about 3*s*. the hundred. /

The amount of the various items is thus:

	£	*s.*	*d.*
1. Hair	0	0	1¼
2. Skin	0	9	0
3. Blood	0	1	9
4. Shoes and nails	0	0	2½
5. Hoofs	0	1	5
6. Fat	0	4	0
7. Flesh	1	8	0
8. Tendons	0	0	3
9. Bones	0	2	0
10. Intestines			
11. Maggots	0	1	5
12. Rat skins			
	2	8	1¾

The dead horse, therefore, which can be purchased at from 8*s*. 6*d*. to 12*s*., produces from £2 9*s*. to £4 14*s*.

needles cross each other (unless, which is exceedingly improbable, they happen to be precisely balanced), they will, when they fall on the bottom of / the tray, tend to place themselves side by side, and the hollow form of the tray assists this disposition. As they have no projection in any part to impede this tendency, or to entangle each other, they are, by continually shaking, arranged lengthwise, in three or four minutes. The direction of the shake is now changed, the needles are but little thrown up, but the tray is shaken endways; the result of which is, that in a minute or two the needles which were previously arranged endways become heaped up in a wall, with their ends against the extremity of the tray. They are then removed, by hundreds at a time, with a broad iron spatula, on which they are retained by the forefinger of the left hand. As this parallel arrangement of the needles must be repeated many times, if a cheap and expeditious method had not been devised, the expense of the manufacture would have been considerably enhanced.

12. Another process in the art of making needles furnishes an example of one of the simplest contrivances which can come under the denomination of a *tool*. After the needles have been arranged in the manner just described, it is necessary to separate them into two parcels, in order that their points may be all in one direction. This is usually done by women and children. The needles are placed sideways in a heap, on a table, in front of each operator, just as they are arranged by the process above described. From five to ten are rolled towards this person with the forefinger of the left hand; this separates them a very small space from each other, and each in its turn is pushed lengthwise to the right or to the left, according to the direction of the point. / This is the usual process, and in it every needle passes individually under the finger of the operator. A small alteration expedites the process considerably: the child puts on the forefinger of its right hand a small cloth cap or fingerstall, and rolling out of the heap from six to twelve needles, he keeps them down by the forefinger of the left hand, whilst he presses the forefinger of the right hand gently against their ends: those which have the points towards the right hand stick into the fingerstall; and the child, removing the finger of the left hand, slightly raises the needles sticking into the cloth, and then pushes them towards the left side. Those needles which had their eyes on the right hand do not stick into the finger cover, and are pushed to the heap on the right side before the repetition of this process. By means of this

simple contrivance each movement of the finger, from one side to the other, carries five or six needles to their proper heap; whereas, in the former method, frequently only one was moved, and rarely more than two or three were transported at one movement to their place.

13. Various operations occur in the arts in which the assistance of an additional hand would be a great convenience to the workman, and in these cases tools or machines of the simplest structure come to our aid: vices of different forms, in which the material to be wrought is firmly grasped by screws, are of this kind, and are used in almost every workshop; but a more striking example may be found in the trade of the nail-maker.

Some kinds of nails, such as those used for defending the soles of coarse shoes, called hobnails, require / a particular form of the head, which is made by the stroke of a die. The workman holds one end of the rod of iron out of which he forms the nails in his left hand; with his right hand he hammers the red-hot end of it into a point, and cutting the proper length almost off, bends it nearly at a right angle. He puts this into a hole in a small stake-iron immediately under a hammer which is connected with a treadle, and has a die sunk in its surface corresponding to the intended form of the head; and having given one part of the form to the head with the small hammer in his hand, he moves the treadle with his foot, disengages the other hammer, and completes the figure of the head; the returning stroke produced by the movement of the treadle striking the finished nail out of the hole in which it was retained. Without this substitution of his foot for another hand, the workman would, probably, be obliged to heat the nails twice over.

14. Another, though fortunately a less general substitution of tools for human hands, is used to assist the labour of those who are deprived by nature, or by accident, of some of their limbs. Those who have had an opportunity of examining the beautiful contrivances for the manufacture of shoes by machinery, which we owe to the fertile invention of Mr Brunel, must have noticed many instances in which the workmen were enabled to execute their task with precision, although labouring under the disadvantages of the loss of an arm or leg. A similar instance occurs at Liverpool, in the Institution for the Blind, where a machine is used by those afflicted with blindness, for weaving sash-lines; it is said to have been / the invention of a person suffering under that calamity. Other examples

might be mentioned of contrivances for the use, the amusement, or the instruction of the wealthier classes, who labour under the same natural disadvantages. These triumphs of skill and ingenuity deserve a double portion of our admiration when applied to mitigate the severity of natural or accidental misfortune; when they supply the rich with occupation and knowledge; when they relieve the poor from the additional evils of poverty and want.

15. *Division of the objects of machinery.* There exists a natural, although, in point of number, a very unequal division amongst machines: they may be classed as; first, *those which are employed to produce power*; and as, secondly, *those which are intended merely to transmit force and execute work.* The first of these divisions is of great importance, and is very limited in the variety of its species, although some of those species consist of numerous individuals.

Of that class of mechanical agents by which motion is transmitted – the lever, the pulley, the wedge, and many others – it has been demonstrated, that no power is gained by their use, however combined. *Whatever force is applied at one point can only be exerted at some other, diminished by friction and other incidental causes*; and it has been further proved, that *whatever is gained in the rapidity of execution is compensated by the necessity of exerting additional force.* These two principles, long since placed beyond the reach of doubt, cannot be too constantly borne in mind. But in limiting our attempts to things which are possible, we are still, as we hope to show, possessed of a field of inexhaustible research, and of / advantages derived from mechanical skill, which have but just begun to exercise their influence on our arts, and may be pursued without limit – contributing to the improvement, the wealth, and the happiness of our race.

16. Of those machines by which we produce power, it may be observed, that although they are to us immense acquisitions, yet in regard to two of the sources of this power – the force of wind and of water – we merely make use of bodies in a state of motion by nature; we change the directions of their movement in order to render them subservient to our purposes, but we neither add to nor diminish the quantity of motion in existence. When we expose the sails of a windmill obliquely to the gale, we check the velocity of a small portion of the atmosphere, and convert its own rectilinear motion into one of rotation in the sails; we thus change the direction of force, but we create no power. The same may be observed with

regard to the sails of a vessel; the quantity of motion given by them is precisely the same as that which is destroyed in the atmosphere. If we avail ourselves of a descending stream to turn a water-wheel, we are appropriating a power which nature may appear, at first sight, to be uselessly and irrecoverably wasting, but which, upon due examination, we shall find she is ever regaining by other processes. The fluid which is falling from a higher to a lower level, carries with it the velocity due to its revolution with the earth at a greater distance from its centre. It will therefore accelerate, although to an almost infinitesimal extent, the earth's daily rotation. The sum of all these increments of / velocity, arising from the descent of all the falling waters on the earth's surface, would in time become perceptible, did not nature, by the process of evaporation, convey the waters back to their sources; and thus again, by removing matter to a greater distance from the centre, destroy the velocity generated by its previous approach.

17. The force of vapour is another fertile source of moving power; but even in this case it cannot be maintained that power is created. Water is converted into elastic vapour by the combustion of fuel. The chemical changes which thus take place are constantly increasing the atmosphere by large quantities of carbonic acid and other gases noxious to animal life. The means by which nature decomposes these elements, or reconverts them into a solid form, are not sufficiently known: but if the end could be accomplished by mechanical force, it is almost certain that the power necessary to produce it would at least equal that which was generated by the original combustion. Man, therefore, does not create power; but, availing himself of his knowledge of nature's mysteries, he applies his talents to diverting a small and limited portion of her energies to his own wants: and, whether he employs the regulated action of steam, or the more rapid and tremendous effects of gunpowder, he is only producing on a small scale compositions and decompositions which nature is incessantly at work in reversing, for the restoration of that equilibrium which we cannot doubt is constantly maintained throughout even the remotest limits of our system. The operations of man participate in the character of their author; they are diminutive, but / energetic during the short period of their existence: whilst those of nature, acting over vast spaces, and unlimited by time, are ever pursuing their silent and resistless career.

18. In stating the broad principle, that all combinations of mechanical

art can only augment the force communicated to the machine at the expense of the time employed in producing the effect, it might, perhaps, be imagined, that the assistance derived from such contrivances is small. This is, however, by no means the case: since the almost unlimited variety they afford, enables us to exert to the greatest advantage whatever force we employ. There is, it is true, a limit beyond which it is impossible to reduce the power necessary to produce any given effect, but it very seldom happens that the methods first employed at all approach that limit. In dividing the knotted root of a tree for fuel, how very different will be the time consumed, according to the nature of the tool made use of! The hatchet, or the adze, will divide it into small parts, but will consume a large portion of the workman's time. The saw will answer the same purpose more quickly and more effectually. This, in its turn, is superseded by the wedge, which rends it in a still shorter time. If the circumstances are favourable, and the workman skilful, the time and expense may be still further reduced by the use of a small quantity of gunpowder exploded in holes judiciously placed in the block.

19. When a mass of matter is to be removed a certain force must be expended; and upon the proper economy of this force the price of transport will depend. A country must, however, have reached / a high degree of civilization before it will have approached the limit of this economy. The cotton of Java is conveyed in junks to the coast of China; but from the seed not being previously separated, three-quarters of the weight thus carried is not cotton. This might, perhaps, be justified in Java by the want of machinery to separate the seed, or by the relative cost of the operation in the two countries. But the cotton itself, as packed by the Chinese, occupies three times the bulk of an equal quantity shipped by Europeans for their own markets. Thus the freight of a given quantity of cotton costs the Chinese nearly twelve times the price to which, by a proper attention to mechanical methods, it might be reduced.* /

* Crawfurd's *Indian Archipelago*.

[21/2]

CHAPTER II

ACCUMULATING POWER

20. Whenever the work to be done requires more force for its execution than can be generated in the time necessary for its completion, recourse must be had to some mechanical method of preserving and condensing a part of the power exerted previously to the commencement of the process. This is most frequently accomplished by a fly-wheel, which is in fact nothing more than a wheel having a very heavy rim, so that the greater part of its weight is near the circumference. It requires great power applied for some time to put this into rapid motion; but when moving with considerable velocity, the effects are exceedingly powerful, if its force be concentrated upon a small object. In some of the iron works where the power of the steam-engine is a little too small for the rollers which it drives, it is usual to set the engine at work a short time before the red-hot iron is ready to be removed from the furnace to the rollers, and to allow it to work with great rapidity until the fly has acquired a velocity rather alarming to those unused to such establishments. On passing the softened mass of iron through the first groove, the engine receives a great and very perceptible check; and its speed is diminished at the next and at each succeeding passage, until the iron bar is reduced to such a size that the ordinary power of the engine is sufficient to roll it. /

21. The powerful effect of a large flywheel when its force can be concentrated on a point, was curiously illustrated at one of the largest of our manufactories. The proprietor was showing to a friend the method of punching holes in iron plates for the boilers of steam-engines. He held in his hand a piece of sheet-iron three-eighths of an inch thick, which he placed under the punch. Observing, after several holes had been made, that the punch made its perforations more and more slowly, he called to the engine-man to know what made the engine work so sluggishly, when it was found that the flywheel and

punching apparatus had been detached from the steam-engine just at the commencement of his experiment.

22. Another mode of accumulating power arises from lifting a weight and then allowing it to fall. A man, even with a heavy hammer, might strike repeated blows upon the head of a pile without producing any effect. But if he raises a much heavier hammer to a much greater height, its fall, though far less frequently repeated, will produce the desired effect.

When a small blow is given to a large mass of matter, as to a pile, the imperfect elasticity of the material causes a small loss of momentum in the transmission of the motion from each particle to the succeeding one; and, therefore, it may happen that the whole force communicated shall be destroyed before it reaches the opposite extremity.

23. The power accumulated within a small space by gunpowder is well known; and, though not strictly an illustration of the subject discussed in this chapter, some of its effects, under peculiar circumstances, are / so singular, that an attempt to explain them may perhaps be excused. If a gun is loaded with ball it will not kick so much as when loaded with small shot; and amongst different kinds of shot, that which is the smallest, causes the greatest recoil against the shoulder. A gun loaded with a quantity of sand, equal in weight to a charge of snipe-shot, kicks still more. If, in loading, a space is left between the wadding and the charge, the gun either recoils violently, or bursts. If the muzzle of a gun has accidentally been stuck into the ground, so as to be stopped up with clay, or even with snow, or if it be fired with its muzzle plunged into water, the almost certain result is that it bursts.

The ultimate cause of these apparently inconsistent effects is, that every force requires *time* to produce its effect; and if the *time* requisite for the elastic vapour within to force out the sides of the barrel, is less than that in which the condensation of the air near the wadding is conveyed in sufficient force to drive the impediment from the muzzle, then the barrel must burst. If sometimes happens that these two forces are so nearly balanced that the barrel only swells; the obstacle giving way before the gun is actually burst.

The correctness of this explanation will appear by tracing step by step the circumstances which arise on discharging a gun loaded with powder confined by a cylindrical piece of wadding, and having its muzzle filled with clay, or some other substance having a moderate degree of resistance. In this case the first effect of the explosion is to produce an enormous pressure on everything confining it, and / to advance the wadding through a very small space. Here let us consider it as at rest for a moment, and examine its condition. The portion of air in immediate contact with the wadding is condensed; and if the wadding were to remain at rest, the air throughout the tube would soon acquire a uniform density. But this would require a small interval of time; for the condensation next the wadding would travel with the velocity of sound to the other end, from whence, being reflected back, a series of waves would be generated, which, aided by the friction of the tube, would ultimately destroy the motion.

But until the first wave reaches the impediment at the muzzle, the air can exert no pressure against it. Now if the velocity communicated to the wadding is very much greater than that of sound, the condensation of the air immediately in advance of it may be very great before the resistance transmitted to the muzzle is at all considerable; in which case the mutual repulsion of the particles of air so compressed, will offer an absolute barrier to the advance of the wadding.*

If this explanation be correct, the additional recoil, when a gun is loaded with small shot or sand, may arise in some measure from the condensation of the air contained between their particles; but chiefly from the velocity communicated by the explosion to those particles of the substances in immediate contact with the powder being greater than that with which a wave can be transmitted through them. It also affords a reason for the success of a method of blasting rocks by filling the upper part of the hole above the powder with sand, instead of / clay rammed hard. That the destruction of the gun barrel does not arise from the property possessed by fluids, and in some measure also by sand and small shot, of pressing equally in all directions, and thus exerting a force against a large portion of the interior surface, seems to be proved by a circumstance mentioned by Le Vaillant and other travellers, that, for the purpose of taking birds without injuring their plumage, they filled the barrel of their fowling pieces with water, instead of loading them with a charge of shot.

24. The same reasoning explains a curious phenomenon which occurs in firing a still more powerfully explosive substance. If we put a small quantity of fulminating silver upon the face of an anvil, and strike it

* See Poisson's remarks, Ecole Polytec. Cahier, xxi, p. 191.

slightly with a hammer, it explodes; but instead of breaking either the hammer or the anvil, it is found that that part of the face of each in contact with the fulminating silver is damaged. In this case the velocity communicated by the elastic matter disengaged may be greater than the velocity of a wave traversing steel; so that the particles at the surface are driven by the explosion so near to those next adjacent, that when the compelling force is removed, the repulsion of the particles within the mass drives back those nearer to the surface, with such force, that they pass beyond the limits of attraction, and are separated in the shape of powder.

25.*i* The success of the experiment of firing a tallow candle through a deal board, would be explained in the same manner, by supposing the velocity of a wave propagated through deal to be greater than that of a wave passing through tallow. /

25.*ii* The boiler of a steam-engine sometimes bursts even during the escape of steam through the safety-valve. If the water in the boiler is thrown upon any part which happens to be red hot, the steam formed in the immediate neighbourhood of that part expands with greater velocity than that with which a wave can be transmitted through the less heated steam; consequently one particle is urged against the next, and an almost invincible obstacle is formed, in the same manner as described in the case of the discharge of a gun. If the safety-valve is closed, it may retain the pressure thus created for a short time, and even when it is open the escape may not be sufficiently rapid to remove all impediment; there may therefore exist momentarily within the boiler pressures of various force, varying from that which can just lift the safety-valve up to that which is sufficient, if exerted during an extremely small space of time, to tear open the boiler itself.

26.This reasoning ought, however, to be admitted with caution; and perhaps some inducement to examine it carefully may be presented by tracing it to extreme cases. It would seem, but this is not a necessary consequence, that a gun might be made so long, that it would burst although no obstacle filled up its muzzle. It should also follow that if, after the gun is charged, the air were extracted from the barrel, though the muzzle be then left closed, the gun ought not to burst. It would also seem to follow from the principle of the explanation, that a body might be projected in air, or other elastic resisting medium, with such force that, after advancing a very short space, it should return in the same direction in which it was projected. /

[27/8]

CHAPTER III

REGULATING POWER

27. Uniformity and steadiness in the rate at which machinery works, are essential both for its effect and its duration. The first illustration which presents itself is that beautiful contrivance, the governor of the steam-engine; which must immediately occur to all who are familiar with that admirable engine. Wherever the increased speed of the engine would lead to injurious or dangerous consequences, this is applied; and it is equally the regulator of the water-wheel which drives a spinning-jenny, or of the windmills which drain our fens. In the dockyard at Chatham, the descending motion of a large platform, on which timber is raised, is regulated by a governor; but as the weight is very considerable, the velocity of this governor is still further checked by causing its motion to take place in water.

28. Another very beautiful contrivance for regulating the number of strokes made by a steam-engine, is used in Cornwall: it is called the *cataract*, and depends on the time required to fill a vessel plunged in water, the opening of the valve through which the fluid is admitted being adjustable at the will of the engine-man.

29. The regularity of the supply of fuel to the fire under the boilers of steam-engines is another mode / of contributing to the uniformity of their rate, and also economizes the consumption of coal. Several patents have been taken out for methods of regulating this supply: the general principle being to make the engine supply the fire with small quantities of fuel at regular intervals by means of a hopper, and to make it diminish this supply when the engine works too quickly. One of the incidental advantages of this plan is, that by throwing on a very small quantity of coal at a time, the smoke is almost entirely consumed. The dampers of ashpits and chimneys are also, in some cases, connected with machines in order to regulate their speed.

30. Another contrivance for regulating the effect of machinery consists in a vane or fly, of little weight, but presenting a large

surface. This revolves rapidly, and soon acquires a uniform rate, which it cannot greatly exceed, because any addition to its velocity produces a much greater addition to the resistance it meets with from the air. The interval between the strokes on the bell of a clock is regulated in this way, and the fly is so contrived, that the interval may be altered by presenting the arms of it more or less obliquely to the direction in which they move. This kind of fly, or vane, is generally used in the smaller kinds of mechanism, and, unlike the heavy fly, it is a destroyer instead of a preserver of force. It is the regulator used in musical boxes, and in almost all mechanical toys.

31. The action of a fly, or vane, suggests the principle of an instrument for measuring the altitude of mountains, which perhaps deserves a trial, since, if it succeed only tolerably, it will form a much more / portable instrument than the barometer. It is well known that the barometer indicates the weight of a column of the atmosphere above it, whose base is equal to the bore of the tube. It is also known that the density of the air adjacent to the instrument will depend both on the weight of air above it, and on the heat of the air at that place. If, therefore, we can measure the density of the air, and its temperature, the height of a column of mercury which it would support in the barometer can be found by calculation. Now the thermometer gives information respecting the temperature of the air immediately; and its density might be ascertained by means of a watch and a small instrument, in which the number of turns made by a vane moved by a constant force, should be registered. The less dense the air in which the vane revolves, the greater will be the number of its revolutions in a given time: and tables could be formed from experiments in partially exhausted vessels, aided by calculation, from which, if the temperature of the air, and the number of revolutions of the vane are given, the corresponding height of the barometer might be found.* /

* To persons who may be inclined to experiment upon this or any other instrument, I would beg to suggest the perusal of the section 'On the Art of Observing', *Observations on the Decline of Science in England*, p. 170. Fellowes, 1828. <*Works of Babbage*, Vol. 7.>

[30/1]

CHAPTER IV

INCREASE AND DIMINUTION OF VELOCITY

32. The fatigue produced on the muscles of the human frame does not altogether depend on the actual force employed in each effort, but partly on the frequency with which it is exerted. The exertion necessary to accomplish every operation consists of two parts: one of these is the expenditure of force which is necessary to drive the tool or instrument; and the other is the effort required for the motion of some limb of the animal producing the action. In driving a nail into a piece of wood, one of these is lifting the hammer, and *propelling* its head against the nail; the other is, *raising* the arm itself, and moving it in order to use the hammer. If the weight of the hammer is considerable, the former part will cause the greatest portion of the exertion. If the hammer is light, the exertion of *raising* the arm will produce the greatest part of the fatigue. It does therefore happen, that operations requiring very trifling force, if frequently repeated, will tire more effectually than more laborious work. There is also a degree of rapidity beyond which the action of the muscles cannot be pressed.

33. The most advantageous load for a porter who carries wood up stairs on his shoulders, has been investigated by M. Coulomb; but he found from / experiment that a man walking up stairs without any load, and raising his burden by means of his own weight in descending, could do as much work in one day, as four men employed in the ordinary way with the most favourable load.

34. The proportion between the velocity with which men or animals move, and the weights they carry, is a matter of considerable importance, particularly in military affairs. It is also of great importance for the economy of labour, to adjust the weight of that part of the animal's body which is moved, the weight of the tool it urges, and the frequency of repetition of these efforts, so as to produce the greatest effect. An instance of the saving of time by

making the same motion of the arm execute two operations instead of one, occurs in the simple art of making the tags of bootlaces: these tags are formed out of very thin, tinned, sheet-iron, and were formerly cut out of long strips of that material into pieces of such a breadth that when bent round they just enclosed the lace. Two pieces of steel have recently been fixed to the side of the shears, by which each piece of tinned-iron as soon as it is cut is bent into a semi-cylindrical form. The additional power required for this operation is almost imperceptible; and it is executed by the same motion of the arm which produces the cut. The work is usually performed by women and children; and with the improved tool more than three times the quantity of tags is produced in a given time.*

35. Whenever the work is itself light, it becomes necessary, in order to economize time, to increase the / velocity. Twisting the fibres of wool by the fingers would be a most tedious operation: in the common spinning-wheel the velocity of the foot is moderate, but by a very simple contrivance that of the thread is most rapid. A piece of catgut passing round a large wheel, and then round a small spindle, effects this change. This contrivance is common to a multitude of machines, some of them very simple. In large shops for the retail of ribands, it is necessary at short intervals to 'take stock', that is, to measure and rewind every piece of riband, an operation which, even with this mode of shortening it, is sufficiently tiresome, but without it would be almost impossible from its expense. The small balls of sewing cotton, so cheap and so beautifully wound, are formed by a machine on the same principle, and but a few steps more complicated.

36. In turning from the smaller instruments in frequent use to the larger and more important machines, the economy arising from the increase of velocity becomes more striking. In converting cast into wrought-iron, a mass of metal, of about a hundredweight, is heated almost to white heat, and placed under a heavy hammer moved by water or steam power. This is raised by a projection on a revolving axis; and if the hammer derived its momentum only from the space through which it fell, it would require a considerably greater time to give a blow. But as it is important that the softened mass of red-hot iron should receive as many blows as possible before it cools, the form of the cam or projection on the axis is such, that the hammer,

* See *Transactions of the Society of Arts*, 1826.

instead of being lifted to a small height, is thrown up / with a jerk, and almost the instant after it strikes against a large beam, which acts as a powerful spring, and drives it down on the iron with such velocity that by these means about double the number of strokes can be made in a given time. In the smaller tilt-hammers, this is carried still further: by striking the tail of the tilt-hammer forcibly against a small steel anvil, it rebounds with such velocity, that from three to five hundred strokes are made in a minute. In the manufacture of anchors, an art in which a similar contrivance is of still greater importance, it has only been recently applied.

37. In the manufacture of scythes, the length of the blade renders it necessary that the workman should move readily, so as to bring every part of it on the anvil in quick succession. This is effected by placing him in a seat suspended by ropes from the ceiling: so that he is enabled, with little bodily exertion, to vary his distance, by pressing his feet against the block which supports the anvil, or against the floor.

38. An increase of velocity is sometimes necessary to render operations possible: thus a person may skate with great rapidity over ice which would not support his weight if he moved over it more slowly. This arises from the fact, that time is requisite for producing the fracture of the ice: as soon as the weight of the skater begins to act on any point, the ice, supported by the water, bends slowly under him; but if the skater's velocity is considerable, he has passed off from the spot which was loaded before the bending has reached the point which would cause the ice to break. /

39. An effect not very different from this might take place if very great velocity were communicated to boats. Let us suppose a flat-bottomed boat, whose bow forms an inclined plane with the bottom, at rest in still water. If we imagine some very great force suddenly to propel this boat, the inclination of the plane at the forepart would cause it to rise in the water; and if the force were excessive, it might even rise out of the water, and advance, by a series of leaps, like a piece of slate or an oyster shell, thrown as a '*duck and drake*'.

If the force were not sufficient to pull the boat out of the water, but were just enough to bring its bottom to the surface, it would be carried along with a kind of gliding motion with great rapidity; for at every point of its course it would require a certain time before it could sink to its usual draft of water; but before that time had elapsed, it would have advanced to another point, and consequently have been raised by the reaction of the water on the inclined plane at its fore-part.

40. The same fact, that bodies moving with great velocity have not time to exert the full effect of their weight, seems to explain a circumstance which appears to be very unaccountable. It sometimes happens that when foot-passengers are knocked down by carriages, the wheels pass over them with scarcely any injury, though, if the weight of the carriage had rested on their body, even for a few seconds, it would have crushed them to death. If the view above taken is correct, the injury in such circumstances will chiefly happen to that part of the body which is struck by the advancing wheel. /

41. An operation in which rapidity is of essential importance is in bringing the produce of mines up to the surface. The shafts through which the produce is raised are sunk at a very great expense, and it is, of course, desirable to sink as few of them as possible. The matter to be extracted is therefore raised by steam-engines with considerable velocity; and without this many of our mines could not be worked with profit.

42. The effect of great velocity in modifying the form of a cohesive substance is beautifully shown in the process for making window-glass, termed '*flashing*', which is one of the most striking operations in our domestic arts. A workman having dipped his iron tube into the glass pot, and loaded it with several pounds of the melted '*metal*', blows out a large globe, which is connected with his rod by a short thick hollow neck. Another workman now fixes to the globe immediately opposite to its neck, an iron rod, the extremity of which has been dipped in the melted glass; and when this is firmly attached, a few drops of water separate the neck of the globe from the iron tube. The rod with the globe attached to it is now held at the mouth of a glowing furnace: and by turning the rod the globe is made to revolve slowly, so as to be uniformly exposed to the heat: the first effect of this softening is to make the glass contract upon itself and to enlarge the opening of the neck. As the softening proceeds, the globe is turned more quickly on its axis, and when very soft and almost incandescent, it is removed from the fire, and the velocity of rotation being still continually increased, the opening enlarges from the effect of the / centrifugal force, at first gradually, until at last the mouth suddenly expands or '*flashes*' out into one large circular sheet of red-hot glass. The neck of the original globe, which is to become the outer part of the sheet, is left thick to admit of this expansion, and

forms the edge of the circular plate of glass, which is called a '*Table*'. The centre presents the appearance of a thick boss or prominence, called the '*Bull's-eye*', at the part by which it was attached to the iron rod.

43. The most frequent reason for employing contrivances for diminishing velocity, arises from the necessity of overcoming great resistances with small power. Systems of pulleys, the crane, and many other illustrations, might also be adduced here as examples; but they belong more appropriately to some of the other causes which we have assigned for the advantages of machinery. The common smoke-jack is an instrument in which the velocity communicated is too great for the purpose required, and it is transmitted through wheels which reduce it to a more moderate rate.

44. Telegraphs are machines for conveying information over extensive lines with great rapidity. They have generally been established for the purposes of transmitting information during war, but the increasing wants of man will probably soon render them subservient to more peaceful objects.

A few years since the telegraph conveyed to Paris information of the discovery of a comet, by M. Gambart, at Marseilles: the message arrived during a sitting of the French Board of Longitude, and was sent in a note from the Minister of the Interior / to Laplace, the President, who received it whilst the writer of these lines was sitting by his side. The object in this instance was, to give the earliest publicity to the fact, and to assure to M. Gambart the title of its first discoverer.

At Liverpool a system of signals is established for the purposes of commerce, so that each merchant can communicate with his own vessel long before she arrives in the port. /

[38/9]

CHAPTER V

EXTENDING THE TIME OF ACTION OF FORCES

45. This is one of the most common and most useful of the employments of machinery. The half minute which we daily devote to the winding-up of our watches is an exertion of labour almost insensible; yet, by the aid of a few wheels, its effect is spread over the whole twenty-four hours. In our clocks, this extension of the time of action of the original force impressed is carried still further; the better kind usually require winding up once in eight days, and some are occasionally made to continue in action during a month, or even a year. Another familiar illustration may be noticed in our domestic furniture: the common jack by which our meat is roasted, is a contrivance to enable the cook in a few minutes to exert a force which the machine retails out during the succeeding hour in turning the loaded spit; thus enabling her to bestow her undivided attention on the other important duties of her vocation. A great number of automatons and mechanical toys moved by springs, may be classed under this division.

46. A small moving power, in the shape of a jack or a spring with a train of wheels, is often of great convenience to the experimental philosopher, and has been used with advantage in magnetic and electric / experiments where the rotation of a disk of metal or other body is necessary, thus allowing to the enquirer the unimpeded use of both his hands. A vane connected by a train of wheels, and set in motion by a heavy weight, has also, on some occasions, been employed in chemical processes, to keep a solution in a state of agitation. Another object to which a similar apparatus may be applied, is the polishing of small specimens of minerals for optical experiments. /

[40/1]

CHAPTER VI

SAVING TIME IN NATURAL OPERATIONS

47. The process of tanning will furnish us with a striking illustration of the power of machinery in accelerating certain processes in which natural operations have a principal effect. The object of this art is to combine a certain principle called *tanning* with every particle of the skin to be tanned. This, in the ordinary process, is accomplished by allowing the skins to soak in pits containing a solution of tanning matter: they remain in the pits six, twelve, or eighteen months; and in some instances (if the hides are very thick), they are exposed to the operation for two years, or even during a longer period. This length of time is apparently required in order to allow the tanning matter to penetrate into the interior of a thick hide. The improved process consists in placing the hides with the solution of tan in close vessels, and then exhausting the air. The effect is to withdraw any air which may be contained in the pores of the hides, and to aid capillary attraction by the pressure of the atmosphere in forcing the tan into the interior of the skins. The effect of the additional force thus brought into action can be equal only to one atmosphere, but a further improvement has been made: the vessel containing the hides is, after exhaustion, filled up with a solution of tan; a small additional quantity is then injected with a forcing-pump. / By these means any degree of pressure may be given which the containing vessel is capable of supporting; and it has been found that, by employing such a method, the thickest hides may be tanned in six weeks or two months.

48. The same process of injection might be applied to impregnate timber with tar, or any other substance capable of preserving it from decay; and if it were not too expensive, the deal floors of houses might thus be impregnated with alumine or other substances, which would render them much less liable to be accidentally set on fire. In some cases it might be useful to impregnate woods with resins,

varnish, or oil; and wood saturated with oil might, in some instances, be usefully employed in machinery for giving a constant, but very minute supply of that fluid to iron or steel, against which it is worked. Some idea of the quantity of matter which can be injected into wood by great pressure, may be formed, from considering the fact stated by Mr Scoresby, respecting an accident which occurred to a boat of one of our whaling-ships. The harpoon having been struck into the fish, the whale in this instance, dived directly down, and carried the boat along with him. On returning to the surface the animal was killed, but the boat, instead of rising, was found suspended beneath the whale by the rope of the harpoon; and on drawing it up, every part of the wood was found to be so completely saturated with water as to sink immediately to the bottom.

49. The operation of bleaching linen in the open air is one for which considerable time is necessary; and although it does not require much labour, yet, from the risk of damage and of robbery from long /exposure, a mode of shortening the process was highly desirable. The method now practised, although not mechanical, is such a remarkable instance of the application of science to the practical purposes of manufactures, that in mentioning the advantages derived from shortening natural operations, it would have been scarcely pardonable to have omitted all allusion to the beautiful application of chlorine, in combination with lime, to the art of bleaching.

50. Another instance more strictly mechanical occurs in some countries where fuel is expensive, and the heat of the sun is not sufficient to evaporate the water from brine springs. The water is first pumped up to a reservoir, and then allowed to fall in small streams through faggots. Thus it becomes divided; and, presenting a large surface, evaporation is facilitated, and the brine which is collected in the vessels below the faggots is stronger than that which was pumped up. After thus getting rid of a large part of the water, the remaining portion is driven off by boiling. The success of this process depends on the condition of the atmosphere with respect to moisture. If the air, at the time the brine falls through the faggots, holds in solution as much moisture as it can contain in an invisible state, no more can be absorbed from the salt water, and the labour expended in pumping is entirely wasted. The state of the air, as to dryness, is therefore an important consideration in fixing the time

when this operation is to be performed; and an attentive examination of its state, by means of the hygrometer, might be productive of some economy of labour.

51. In some countries, where wood is scarce, the evaporation of salt water is carried on by a large / collection of ropes which are stretched perpendicularly. In passing down the ropes, the water deposits the sulphate of lime which it held in solution, and gradually incrusts them, so that in the course of twenty years, when they are nearly rotten, they are still sustained by the surrounding incrustation, thus presenting the appearance of a vast collection of small columns.

52. Amongst natural operations perpetually altering the surface of our globe, there are some which it would be advantageous to accelerate. The wearing down of the rocks which impede the rapids of navigable rivers, is one of this class. A very beautiful process for accomplishing this object has been employed in America. A boat is placed at the bottom of the rapid, and kept in its position by a long rope which is firmly fixed on the bank of the river near the top. An axis, having a wheel similar to the paddle-wheel of a steamboat fixed at each end of it, is placed across the boat; so that the two wheels and their connecting axis shall revolve rapidly, being driven by the force of the passing current. Let us now imagine several beams of wood shod with pointed iron fixed at the ends of strong levers, projecting beyond the bow of the boat, as in the annexed representation. /

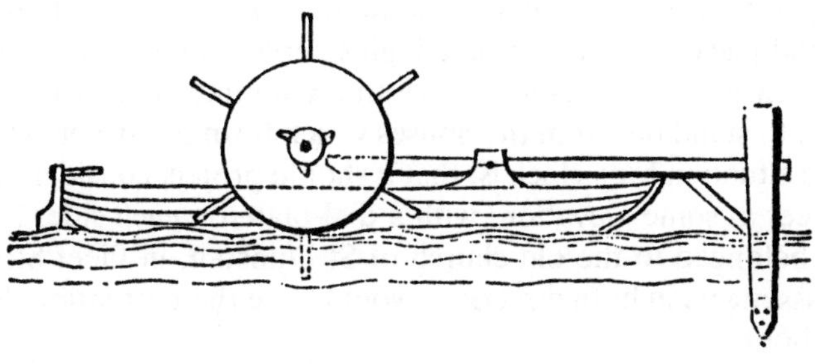

If these levers are at liberty to move up and down, and if one or more projecting pieces, called cams, are fixed on the axis opposite to the end of each lever, the action of the stream upon the wheels will keep up a perpetual succession of blows. The sharp-pointed shoe striking upon the rock at the bottom, will continually detach small pieces, which the stream will immediately carry off. Thus, by the mere action of the river itself, a constant and most effectual system of pounding the rock at its bottom is established. A single workman may, by the aid of a rudder, direct the boat to any required part of the stream; and when it is necessary to move up the rapid, as the channel is cut, he can easily cause the boat to advance by means of a capstan.

53. When the object of the machinery just described has been accomplished, and the channel is sufficiently deep, a slight alteration converts the apparatus to another purpose almost equally advantageous. The stampers and the projecting pieces on the axis are removed, and a barrel of wood or metal, surrounding part of the axis, and capable, at pleasure, of being connected with, or disconnected from the axis itself, is substituted. The rope which hitherto fastened the boat, is now fixed to this barrel; and if the barrel is loose upon the axis, the paddle-wheel makes the axis only revolve, and the boat remains in its place: but the moment the axis is attached to its surrounding barrel, this begins to turn, and winding up the rope, the boat is gradually drawn up against the stream; and may be employed as a kind of tug-boat for vessels which have occasion to ascend the rapid. When the tug-boat reaches the / summit the barrel is released from the axis, and friction being applied to moderate its velocity, the boat is allowed to descend.

54. Clocks occupy a very high place amongst instruments by means of which human time is economized: and their multiplication in conspicuous places in large towns is attended with many advantages. Their position, nevertheless, in London, is often very ill chosen; and the usual place, halfway up on a high steeple, in the midst of narrow streets, in a crowded city, is very unfavourable, unless the church happen to stand out from the houses which form the street. The most eligible situation for a clock is, that it should project considerably into the street at some elevation, with a dial-plate on each side, like that which belonged to the old church of St Dunstan, in Fleet Street, so that passengers in both directions would have their attention directed to the hour.

55. A similar remark applies, with much greater force, to the present defective mode of informing the public of the position of the receiving houses for the twopenny and general post. In the lowest corner of the window of some attractive shop is found a small slit, with a brass plate indicating its important office so obscurely, that it seems to be an object rather to prevent its being conspicuous. No striking sign assists the anxious enquirer, who, as the moments

rapidly pass which precede the hour of closing, torments the passenger with his enquiries for the nearest post-office. He reaches it, perhaps, just as it is closed; and must then either hasten to a distant part of the town in order to procure the admission of his letters, or give up the idea of forwarding / them by that post; and thus, if they are foreign letters, he may lose, perhaps, a week or a fortnight by waiting for the next packet.

The inconvenience in this and in some other cases, is of perpetual and everyday occurrence; and though, in the greater part of the individual cases, it may be of trifling moment, the sum of all these produces an amount, which it is always worthy of the government of a large and active population to attend to. The remedy is simple and obvious: it would only be necessary, at each letter-box, to have a light frame of iron projecting from the house over the pavement, and carrying the letters G. P., or T. P., or any other distinctive sign. All private signs are at present very properly prohibited from projecting into the street: the passenger, therefore, would at once know where to direct his attention, in order to discover a post-office; and those letter-boxes which occurred in the great thoroughfares could not fail to be generally known. /

[47/8]

CHAPTER VII

EXERTING FORCES TOO GREAT FOR HUMAN POWER, AND EXECUTING OPERATIONS TOO DELICATE FOR HUMAN TOUCH

56. It requires some skill and a considerable apparatus to enable many men to exert their whole force at a given point; and when this number amounts to hundreds or to thousands, additional difficulties present themselves. If ten thousand men were hired to act simultaneously, it would be exceedingly difficult to discover whether each exerted his whole force, and consequently, to be assured that each man did the duty for which he was paid. And if still larger bodies of men or animals were necessary, not only would the difficulty of directing them become greater, but the expense would increase from the necessity of transporting food for their subsistence.

The difficulty of enabling a large number of men to exert their force at the same instant of time has been almost obviated by the use of sound. The whistle of the boatswain performs this service on board ships; and in removing, by manual force, the vast mass of granite, weighing above 1,400 tons, on which the equestrian figure of Peter the Great is placed at St Petersburgh, a drummer was always / stationed on its summit to give the signal for the united efforts of the workmen.

An ancient Egyptian drawing was discovered a few years since, by Champollion, in which a multitude of men appeared harnessed to a huge block of stone, on the top of which stood a single individual with his hands raised above his head, apparently in the act of clapping them, for the purpose of insuring the exertion of their combined force at the same moment of time.

57. In mines, it is sometimes necessary to raise or lower great weights by capstans requiring the force of more than one hundred men. These work upon the surface; but the directions must be communicated

from below, perhaps from the depth of two hundred fathoms. This communication, however, is accomplished with ease and certainty by signals: the usual apparatus is a kind of clapper placed on the surface close to the capstan, so that every man may hear, and put in motion from below by a rope passing up the shaft.

At Wheal Friendship mine in Cornwall, a different contrivance is employed: there is in that mine an inclined plane, passing underground about two-thirds of a mile in length. Signals are communicated by a continuous rod of metal, which being struck below, the blow is distinctly heard on the surface.

58. In all our larger manufactories numerous instances occur of the application of the power of steam to overcome resistances which it would require far greater expense to surmount by means of animal labour. The twisting of the largest cables, the rolling, hammering, and cutting large masses of iron, / the draining of our mines, all require enormous exertions of physical force continued for considerable periods of time. Other means are had recourse to when the force required is great, and the space through which it is to act is small. The hydraulic press of Bramah can, by the exertion of one man, produce a pressure of 1,500 atmospheres; and with such an instrument a hollow cylinder of wrought iron three inches thick has been burst. In rivetting together the iron plates, out of which steam-engine boilers are made, it is necessary to produce as close a joint as possible. This is accomplished by using the rivets red-hot: while they are in that state the two plates of iron are rivetted together, and the contraction which the rivet undergoes in cooling draws them together with a force which is only limited by the tenacity of the metal of which the rivet itself is made.

59. It is not alone in the greater operations of the engineer or the manufacturer, that those vast powers which man has called into action, in availing himself of the agency of steam, are fully developed. Wherever the individual operation demanding little force for its own performance is to be multiplied in almost endless repetition, commensurate power is required. It is the same 'giant arm' which twists 'the largest cable', that spins from the cotton plant an 'almost gossamer thread'. Obedient to the hand which called into action its resistless powers, it contends with the ocean and the storm, and rides triumphant through dangers and difficulties unattempted by the older modes of navigation. It is the same engine that, in its more regulated action, weaves the canvas it may one day supersede; or,

with / almost fairy fingers, entwines the meshes of the most delicate fabric that adorns the female form.*

60. The Fifth Report of the Select Committee of the House of Commons on the Holyhead Roads furnishes ample proof of the great superiority of steam vessels. The following extracts are taken from the evidence of Captain Rogers, the commander of one of the packets:

Question. Are you not perfectly satisfied, from the experience you have had, that the steam vessel you command is capable of performing what no sailing vessel can do?

Answer. Yes.

Question. During your passage from Gravesend to the Downs, could any square-rigged vessel, from a first-rate down to a sloop of war, have performed the voyage you did in the time you did it in the steamboat?

Answer. No; it was impossible. In the Downs we passed several Indiamen, and 150 sail there that could not move down the channel; and at the back of Dungeness we passed 120 more.

Question. At the time you performed that voyage, with the weather you have described, from the Downs to Milford, if that weather had continued twelve months, would any square-rigged vessel have performed it?

Answer. They would have been a long time about it; probably, would have been weeks instead of days. A sailing vessel would not have beat up to Milford, as we did, in twelve months.

61. The process of printing on the silver paper, which is necessary for bank-notes, is attended with some inconvenience, from the necessity of damping the paper previously to taking the impression. It was difficult to do this uniformly; and in the old process of dipping a parcel of several sheets together into a vessel of water, the outside sheets becoming much more wet than the others, were very apt to be torn. A method has been adopted at the Bank of Ireland which obviates this inconvenience. The whole quantity of paper to be damped is placed in a close vessel from which the air is exhausted; water is then admitted, and every leaf is completely wetted; the paper is then removed to a press, and all the superfluous moisture is squeezed out.

* The importance and diversified applications of the steam-engine were most ably enforced in the speeches made at a public meeting, held (June, 1824) for the purpose of proposing the erection of a monument to the memory of James Watt; these were subsequently printed.

62. The operation of pulverizing solid substances and of separating the powders of various degrees of fineness, is common in the arts: and as the best graduated sifting fails in effecting this separation with sufficient delicacy, recourse is had to suspension in a fluid medium. The substance when reduced by grinding to the finest powder is agitated in water, which is then drawn off: the coarsest portion of the suspended matter first subsides, and that which requires the longest time to fall down is the finest. In this manner even emery powder, a substance of great density, is separated into the various degrees of fineness which are required. Flints, after being burned and ground, are suspended in water, in order to mix them intimately with clay, which is also suspended in the same fluid, for the formation of porcelain. The water is then in part evaporated by heat, and the plastic compound, out of which our / most beautiful porcelain is formed, remains. It is a curious fact, and one which requires further examination than it has yet received, that, if this mixture be suffered to remain long at rest before it is worked up, it becomes useless; for it is then found that the silex, which at first was uniformly mixed, becomes aggregated together in small lumps. This parallel to the formation of flints in the chalk strata deserves attention.*

63. The slowness with which powders subside, depends partly on the specific gravity of the substance, and partly on the magnitude of the particles themselves. Bodies, in falling through a resisting medium, after a certain time acquire a uniform velocity, which is called their terminal velocity, with which they continue to descend: when the particles are very small, and the medium dense, as water, this terminal velocity is soon arrived at. Some of the finer powders even of emery require several hours to subside through a few feet of water, and the mud pumped up into our cisterns by some of the water companies is suspended during a still longer time. These facts furnish us with some idea of the great extent over which deposits of river mud may be spread; for if the mud of any river whose waters enter the Gulf Stream, sink through one foot in an hour, it might be carried by that stream 1,500 miles before it had sunk to the depth of 600 or 700 feet. /

64. A number of small filaments of cotton project from even the best spun thread, and when this thread is woven into muslin they injure its

* Some observations on this subject, by Dr Fitton, occur in the appendix to Captain King's Survey of the Coast of Australia, vol. ii, p. 397. London, 1826.

appearance. To cut these off separately is quite impossible, but they are easily removed by passing the muslin rapidly over a cylinder of iron kept at a dull red heat: the time during which each portion of the muslin is in contact with the red-hot iron is too short to heat it to the burning point; but the filaments being much finer, and being pressed close to the hot metal, are burnt.

The removal of these filaments from patent net is still more necessary for its perfection. The net is passed at a moderate velocity through a flame of gas issuing from a very long and narrow slit. Immediately above the flame a long funnel is fixed, which is connected with a large air-pump worked by a steam-engine. The flame is thus drawn forcibly through the net, and all the filaments on both sides of it are burned off at one operation. Previously to this application of the air-pump, the net acting in the same way, although not to the same extent, as the wire-gauze in Davy's safety lamp, cooled down the flame so as to prevent the combustion of the filaments on the upper side: the air-pump by quickening the current of ignited gas, removes this inconvenience. /

[54/5]

CHAPTER VIII

REGISTERING OPERATIONS

65. One great advantage which we may derive from machinery is from the check which it affords against the inattention, the idleness, or the dishonesty of human agents. Few occupations are more wearisome than counting a series of repetitions of the same fact; the number of paces we walk affords a tolerably good measure of distance passed over, but the value of this is much enhanced by possessing an instrument, the pedometer, which will count for us the number of steps we have made. A piece of mechanism of this kind is sometimes applied to count the number of turns made by the wheel of a carriage, and thus to indicate the distance travelled: an instrument, similar in its object, but differing in its construction, has been used for counting the number of strokes made by a steam-engine, and the number of coins struck in a press. One of the simplest instruments for counting any series of operations, was contrived by Mr Donkin.*

66. Another instrument for registering is used in some establishments for calendering and embossing. Many hundred thousand yards of calicoes and stuffs undergo these operations weekly; and as the price paid for the process is small, the value of the / time spent in measuring them would bear a considerable proportion to the profit. A machine has, therefore, been contrived for measuring and registering the length of the goods as they pass rapidly through the hands of the operator, by which all chance of erroneous counting is avoided.

67. Perhaps the most useful contrivance of this kind, is one for ascertaining the vigilance of a watchman. It is a piece of mechanism connected with a clock placed in an apartment to which the watchman has not access; but he is ordered to pull a string situated in a certain part of his round once in every hour. The instrument, aptly

* Transactions of the Society of Arts, 1819, p. 116.

called a *tell-tale*, informs the owner whether the man has missed any, and what hours during the night.

68. It is often of great importance, both for regulations of excise as well as for the interest of the proprietor, to know the quantity of spirits or of other liquors which have been drawn off by those persons who are allowed to have access to the vessels during the absence of the inspectors or principals. This may be accomplished by a peculiar kind of stop-cock – which will, at each opening, discharge only a certain measure of fluid – the number of times the cock has been turned being registered by a counting apparatus accessible only to the master.

69. The time and labour consumed in gauging the contents of casks partly filled, has led to an improvement which, by the simplest means, obviates a considerable inconvenience, and enables any person to read off, on a scale, the number of gallons contained in any vessel, as readily as he does the degree of heat indicated by his thermometer. A small stop-cock / connects the bottom of the cask with a glass tube of narrow bore fixed to a scale on the side of the cask, and rising a little above its top. The plug of the cock may be turned into three positions: in the first, it cuts off all communication with the cask: in the second, it opens a communication between the cask and the glass tube: and, in the third, it cuts off the connection between the cask and the tube, and opens a communication between the tube and any vessel held beneath the cock to receive its contents. The scale of the tube is graduated by pouring into the cask successive quantities of water, while the communication between the cask and the tube is open. Lines are then drawn on the scale opposite the places in the tube to which the water rises at each addition, and the scale being thus formed by actual measurement,* the contents of each cask are known by inspection, and the tedious process of gauging is altogether dispensed with. Other advantages accrue from this simple contrivance, in the great economy of time which it introduces in making mixtures of different spirits, in taking stock, and in receiving spirit from the distiller.

70. The gas-meter, by which the quantity of gas used by each consumer is ascertained, is another instrument of this kind. They are of various forms, but all of them intended to register the number of

* This contrivance is due to Mr Heneky, of High Holborn, in whose establishment it is in constant use.

cubic feet of gas which has been delivered. It is very desirable that these meters should be obtainable at a moderate price, and that every consumer should / employ them; because, by making each purchaser pay only for what he consumes, and by preventing that extravagant waste of gas which we frequently observe, the manufacturer of gas will be enabled to make an equal profit at a diminished price to the consumer.

71. The sale of water by the different companies in London, might also, with advantage, be regulated by a meter. If such a system were adopted, much water which is now allowed to run to waste would be saved, and an unjust inequality between the rates charged on different houses by the same company be avoided.

72. Another most important object to which a meter might be applied, would be to register the quantity of water passing into the boilers of steam-engines. Without this, our knowledge of the quantity evaporated by different boilers, and with fireplaces of different constructions, as well as our estimation of the *duty* of steam-engines, must evidently be imperfect.

73. Another purpose to which machinery for registering operations is applied with much advantage is the determination of the average effect of natural or artificial agents. The mean height of the barometer, for example, is ascertained by noting its height at a certain number of intervals during the twenty-four hours. The more these intervals are contracted, the more correctly will the mean be ascertained; but the true mean ought to be influenced by each momentary change which has occurred. Clocks have been proposed and made with this object, by which a sheet of paper is moved, slowly and uniformly, before a pencil fixed to a float upon the surface of the / mercury in the cup of the barometer. Sir David Brewster proposed, several years ago, to suspend a barometer, and swing it as a pendulum. The variations in the atmosphere would thus alter the centre of oscillation, and the comparison of such an instrument with a good clock, would enable us to ascertain the mean altitude of the barometer during any interval of the observer's absence.*

An instrument for measuring and registering the quantity of rain, was invented by Mr John Taylor, and described by him in the

* About seven or eight years since, without being aware of Sir David Brewster's proposal, I adapted a barometer, as a pendulum, to the works of a common eight-day clock; it remained in my library for several months, but I have mislaid the observations which were made.

Philosophical Magazine.[a] It consists of an apparatus in which a vessel that receives the rain falling into the reservoir tilts over as soon as it is full, and then presents another similar vessel to be filled, which in like manner, when full, tilts the former one back again. The number of times these vessels are emptied is registered by a train of wheels; and thus, without the presence of the observer, the quantity of rain falling during a whole year may be measured and recorded.

Instruments might also be contrived to determine the average force of traction of horses – of the wind – of a stream – or of any irregular and fluctuating effort of animal or other natural force.

74. Clocks and watches may be considered as instruments for registering the number of vibrations performed by a pendulum or a balance. The mechanism by which these numbers are counted is technically / called a *scapement.* It is not easy to describe: but the various contrivances which have been adopted for this purpose, are amongst the most interesting and most ingenious to which mechanical science has given birth. Working models, on an enlarged scale, are almost necessary to make their action understood by the unlearned reader; and, unfortunately, these are not often to be met with. A very fine collection of such models exists amongst the collection of instruments at the University of Prague.

Instruments of this kind have been made to extend their action over considerable periods of time, and to register not merely the hour of the day, but the days of the week, of the month, of the year, and also to indicate the occurrence of several astronomical phenomena.

Repeating clocks and watches may be considered as instruments for registering time, which communicate their information only when the owner requires it, by pulling a string, or by some similar application.

An apparatus has recently been applied to watches, by which the hand which indicates seconds leaves a small dot of ink on the dial-plate whenever a certain stop or detent is pushed in. Thus, whilst the eye is attentively fixed on the phenomenon to be observed, the finger registers on the face of the watch-dial the commencement and the end of its appearance.

75. Several instruments have been contrived for awakening the attention of the observer at times previously fixed upon. The various kinds of alarums connected with clocks and watches are of this kind. In

[a] John Taylor, 'Description of a rain gauge', *Philosophical Magazine*, Vol. 2 (1827), pp. 406–8.

some instances it is desirable to be able to set / them so as to give notice at many successive and distant points of time, such as those of the arrival of given stars on the meridian. A clock of this kind is used at the Royal Observatory at Greenwich.

76. An earthquake is a phenomenon of such frequent occurrence, and so interesting, both from its fearful devastations as well as from its connection with geological theories, that it becomes important to possess an instrument which shall, if possible, indicate the direction of the shock, as well as its intensity. An observation made a few years since at Odessa, after an earthquake which happened during the night, suggests a simple instrument by which the direction of the shock may be determined.

A glass vase, partly filled with water, stood on the table of a room in a house at Odessa; and, from the coldness of the glass, the inner part of the vessel above the water was coated with dew. Several very perceptible shocks of an earthquake happened between three and four o'clock in the morning; and when the observer got up, he remarked that the dew was brushed off at two opposite sides of the glass by a wave which the earthquake had caused in the water. The line joining the two highest points of this wave was, of course, that in which the shock travelled. This circumstance, which was accidentally noticed by an engineer at Odessa,* suggests the plan of keeping, in countries subject to earthquakes, glass vessels partly filled with treacle, or some unctuous fluid, so that when any lateral motion is communicated to them from the earth, the / adhesion of the liquid to the glass shall enable the observer, after some interval of time, to determine the direction of the shock.

In order to obtain some measure of the vertical oscillation of the earth, a weight might be attached to a spiral spring, or a pendulum might be sustained in a horizontal position, and a sliding index be moved by either of them, so that the extreme deviations should be indicated by it. This, however, would not give even the comparative measure accurately, because a difference in the velocity of the rising or falling of the earth's surface would affect the instrument. /

* Mémoires de l'Académie des Sciences de Petersburgh, 6e serie, tom. i, p. 4.

[62/3]

CHAPTER IX

ECONOMY OF THE MATERIALS EMPLOYED

77. The precision with which all operations by machinery are executed, and the exact similarity of the articles thus made, produce a degree of economy in the consumption of the raw material which is, in some cases, of great importance. The earliest mode of cutting the trunk of a tree into planks, was by the use of the hatchet or the adze. It might, perhaps, be first split into three or four portions, and then each portion was reduced to a uniform surface by those instruments. With such means the quantity of plank produced would probably not equal the quantity of the raw material wasted by the process: and, if the planks were thin, would certainly fall far short of it. An improved tool, completely reverses the case: in converting a tree into thick planks, the saw causes a waste of a very small fractional part; and even in reducing it to planks of only an inch in thickness, does not waste more than an eighth part of the raw material. When the thickness of the plank is still further reduced, as is the case in cutting wood for veneering, the quantity of material destroyed again begins to bear a considerable proportion to that which is used; and hence circular saws, having a very thin blade, have been employed for such purposes. In order to economize / still further the more valuable woods, Mr Brunel contrived a machine which, by a system of blades, cut off the veneer in a continuous shaving, thus rendering the whole of the piece of timber available.

78. The rapid improvements which have taken place in the printing-press during the last twenty years, afford another instance of saving in the materials consumed, which has been well ascertained by measurement, and is interesting from its connection with literature. In the old method of inking type, by large hemispherical balls stuffed and covered with leather, the printer, after taking a small portion of ink from the ink-block, was continually rolling the balls in various directions against each other, in order that a thin layer of ink might

be uniformly spread over their surface. This he again transferred to the type by a kind of rolling action. In such a process, even admitting considerable skill in the operator, it could not fail to happen that a large quantity of ink should get near the edges of the balls, which, not being transferred to the type, became hard and useless, and was taken off in the form of a thick black crust. Another inconvenience also arose – the quantity of ink spread on the block not being regulated by measure, and the number and direction of the transits of the inking-balls over each other depending on the will of the operator, and being consequently irregular, it was impossible to place on the type a uniform layer of ink, of the quantity exactly sufficient for the impression. The introduction of cylindrical rollers of an elastic substance, formed by the mixture of glue and treacle, superseded the inking-balls, and produced considerable saving in the consumption of / ink: but the most perfect economy was only to be produced by mechanism. When printing-presses, moved by the power of steam, were introduced, the action of these rollers was found to be well adapted to their performance; and a reservoir of ink was formed, from which a roller regularly abstracted a small quantity at each impression. From three to five other rollers spread this portion uniformly over a slab (by most ingenious contrivances varied in almost each kind of press), and another travelling roller, having fed itself on the slab, passed and repassed over the type just before it gave the impression to the paper.

In order to show that this plan of inking puts the proper quantity of ink upon the type, we must prove, first – that the quantity is not too little: this would soon have been discovered from the complaints of the public and the booksellers; and, secondly – that it is not too great. This latter point was satisfactorily established by an experiment. A few hours after one side of a sheet of paper has been printed upon, the ink is sufficiently dry to allow it to receive the impression upon the other; and, as considerable pressure is made use of, the tympan on which the side first printed is laid, is guarded from soiling it by a sheet of paper called the *set-off sheet*. This paper receives, in succession, every sheet of the work to be printed, acquiring from them more or less of the ink, according to their dryness, or the quantity upon them. It was necessary in the former process, after about one hundred impressions, to change this *set-off sheet*, which then became too much soiled for further use. In the new method of printing by machinery, / no such sheet is used, but a blanket is employed as its substitute; this does not require changing above once in five thousand impressions, and instances have occurred of its remaining sufficiently clean for twenty thousand. Here, then, is a proof that the quantity of superfluous ink put upon the paper in machine-printing is so small, that, if multiplied by five thousand, and in some instances even by twenty thousand, it is only sufficient to render useless a single piece of clean cloth.*

The following were the results of an accurate experiment upon the effect of the process just described, made at one of the largest printing establishments in the metropolis.† Two hundred reams of paper were printed off, the old method of inking with balls being employed; two hundred reams of the same paper, and for the same book, were then printed off in the presses which inked their own type. *The consumption of ink by the machine was to that by the balls as four to nine, or rather less than one-half.* /

* In the very best kind of printing, it is necessary, in the old method, to change the set-off sheet once in twelve times. In printing the same kind of work by machinery, the *blanket* is changed once in 2,000.

† This experiment was made at the establishment of Mr Clowes, in Stamford Street.

CHAPTER X

OF THE IDENTITY OF THE WORK WHEN IT IS OF THE SAME KIND, AND ITS ACCURACY WHEN OF DIFFERENT KINDS

79. Nothing is more remarkable, and yet less unexpected, than the perfect identity of things manufactured by the same tool. If the top of a circular box is to be made to fit over the lower part, it may be done in the lathe by gradually advancing the tool of the sliding-rest; the proper degree of tightness between the box and its lid being found by trial. After this adjustment, if a thousand boxes are made, no additional care is required; the tool is always carried up to the stop, and each box will be equally adapted to every lid. The same identity pervades all the arts of printing; the impressions from the same block, or the same copperplate, have a similarity which no labour could produce by hand. The minutest traces are transferred to all the impressions, and no omission can arise from the inattention or unskilfulness of the operator. The steel punch, with which the card-wadding for a fowling-piece is cut, if it once perform its office with accuracy, constantly reproduces the same exact circle.

80. The accuracy with which machinery executes its work is, perhaps, one of its most important advantages: it may, however, be contended, that a / considerable portion of this advantage may be resolved into saving of time; for it generally happens, that any improvement in tools increases the quantity of work done in a given time. Without tools, that is, by the mere efforts of the human hand, there are, undoubtedly, multitudes of things which it would be impossible to make. Add to the human hand the rudest cutting-instrument, and its powers are enlarged: the fabrication of many things then becomes easy, and that of others possible with great labour. Add the saw to the knife or the hatchet, and other works become possible, and a new course of difficult operations is brought into view, whilst many of the former are rendered easy. This

observation is applicable even to the most perfect tools or machines. It would be *possible* for a very skilful workman, with files and polishing substances, to form a cylinder out of a piece of steel; but the time which this would require would be so considerable, and the number of failures would probably be so great, that for all practical purposes such a mode of producing a steel cylinder might be said to be impossible. The same process by the aid of the lathe and the sliding-rest is the everyday employment of hundreds of workmen.

81. Of all the operations of mechanical art, that of turning is the most perfect. If two surfaces are worked against each other, whatever may have been their figure at the commencement, there exists a tendency in them both to become portions of spheres. Either of them may become convex, and the other concave, with various degrees of curvature. A plane surface is the line of separation between convexity and concavity, and is most difficult to hit; it is more / easy to make a good circle than to produce a straight line. A similar difficulty takes place in figuring specula for telescopes; the parabola is the surface which separates the hyperbolic from the elliptic figure, and is the most difficult to form. If a spindle, not cylindrical at its end, be pressed into a hole not circular, and kept constantly turning, there is a tendency in these two bodies so situated to become conical, or to have circular sections. If a triangular-pointed piece of iron be worked round in a circular hole the edges will gradually wear, and it will become conical. These facts, if they do not explain, at least illustrate the principles on which the excellence of work formed in the lathe depends. /

CHAPTER XI

OF COPYING

82. The two last-mentioned sources of excellence in the work produced by machinery depend on a principle which pervades a very large portion of all manufactures, and is one upon which the cheapness of the articles produced seems greatly to depend. The principle alluded to is that of copying, taken in its most extensive sense. Almost unlimited pains are, in some instances, bestowed on the original, from which a series of copies is to be produced; and the larger the number of these copies, the more care and pains can the manufacturer afford to lavish upon the original. It may thus happen, that the instrument or tool actually producing the work, shall cost five or even ten thousand times the price of each individual specimen of its power.

As the system of copying is of so much importance, and of such extensive use in the arts, it will be convenient to classify a considerable number of those processes in which it is employed. The following enumeration however is not offered as a complete list; and the explanations are restricted to the shortest possible detail which is consistent with a due regard to making the subject intelligible. /

Operations of copying are effected under the following circumstances:

by printing from cavities	by stamping
by printing from surface	by punching
by casting	with elongation
by moulding	with altered dimensions

Of printing from cavities

83. The art of printing, in all its numerous departments, is essentially an art of copying. Under its two great divisions, printing from hollow

lines, as in copperplate, and printing from surface, as in block printing, are comprised numerous arts.

84. *Copperplate printing.* In this instance, the copies are made by transferring to paper, by means of pressure, a thick ink, from the hollows and lines cut in the copper. An artist will sometimes exhaust the labour of one or two years upon engraving a plate, which will not, in some cases furnish above five hundred copies in a state of perfection.

85. *Engravings on steel.* This art is like that of engraving on copper, except that the number of copies is far less limited. A bank-note engraved as a copperplate, will not give above three thousand impressons without a sensible deterioration. Two impressions of a bank-note engraved on steel were examined by one of our most eminent artists,* who found it difficult to pronounce with any confidence, which was the earliest impression. One of these was a proof from amongst the first thousand, the other was taken after between seventy and eighty thousand had been printed off. /

86. *Music printing.* Music is usually printed from pewter plates, on which the characters have been impressed by steel punches. The metal being much softer than copper, is liable to scratches, which detain a small portion of the ink. This is the reason of the dirty appearance of printed music. A new process has recently been invented by Mr Cowper, by which this inconvenience will be avoided. The improved method, which give sharpness to the characters, is still an art of copying; but it is effected by surface printing, nearly in the same manner as calico-printing from blocks, to be described hereafter, 96. The method of printing music from pewter plates, although by far the most frequently made use of, is not the only one employed, for music is occasionally printed from stone. Sometimes also it is printed with moveable type; and occasionally the musical characters are printed on the paper, and the lines printed afterwards. Specimens of both these latter modes of music-printing may be seen in the splendid collection of impressions from the types of the press of Bodoni at Parma: but notwithstanding the great care bestowed on the execution of that work, the perpetual interruption of continuity in the lines, arising from the use of moveable types, when the characters and lines are printed at the same time, is apparent.

87. *Calico printing from cylinders.* Many of the patterns on printed

* The late Mr Lowry.

calicos are copies by printing from copper cylinders about four or five inches in diameter, on which the desired pattern has been previously engraved. One portion of the cylinders is exposed to the ink, whilst an elastic scraper of very thin steel, by being pressed forcibly against / another part, removes all superfluous ink from the surface previously to its reaching the cloth. A piece of calico twenty-eight yards in length rolls through this press, and is printed in four or five minutes.

88. *Printing from perforated sheets of metal, or stencilling.* Very thin brass is sometimes perforated in the form of letters, usually those of a name; this is placed on any substance which it is required to mark, and a brush dipped in some paint is passed over the brass. Those parts which are cut away admit the paint, and thus a copy of the name appears on the substance below. This method, which affords rather a coarse copy, is sometimes used for paper with which rooms are covered, and more especially for the borders. If a portion be required to match an old pattern, this is, perhaps the most economical way of producing it.

89. Coloured impressions of leaves upon paper may be made by a kind of surface printing. Such leaves are chosen as have considerable inequalities: the elevated parts of these are covered, by means of an inking ball, with a mixture of some pigment ground up in linseed oil; the leaf is then placed between two sheets of paper, and being gently pressed, the impression from the elevated parts on each side appear on the corresponding sheets of paper.

90. The beautiful red cotton handkerchiefs dyed at Glasgow have their pattern given to them by a process similar to stencilling, except that instead of *printing* from a pattern, the reverse operation – that of *discharging* a part of the colour from a cloth already dyed – is performed. A number of handkerchiefs are pressed with very great force between two plates / of metal, which are similarly perforated with round or lozenge-shaped holes, according to the intended pattern. The upper plate of metal is surrounded by a rim, and a fluid which has the property of discharging the red dye is poured upon that plate. This liquid passes through the holes in the metal, and also through the calico; but, owing to the great pressure opposite all the parts of the plates not cut away, it does not spread itself beyond the pattern. After this, the handkerchiefs are washed, and the pattern of each is a copy of the perforations in the metal-plate used in the process.

Another mode by which a pattern is formed by discharging colour from a previously dyed cloth, is to print on it a pattern with paste; then, passing it into the dying-vat, it comes out dyed of one uniform colour. But the paste has protected the fibres of the cotton from the action of the dye or mordant; and when the cloth so dyed is well washed, the paste is dissolved, and leaves uncoloured all those parts of the cloth to which it was applied.

Printing from surface

91.[a] This second department of printing is of more frequent application in the arts than that which has just been considered.

92. *Printing from wooden blocks.* A block of box wood is, in this instance, the substance out of which the pattern is formed: the design being sketched upon it, the workman cuts away with sharp tools every part except the lines to be represented in the impression. This is exactly the reverse of the process of engraving on copper, in which every line to / be represented is cut away. The ink, instead of filling the cavities cut in the wood, is spread upon the surface which remains, and is thence transferred to the paper.

93. *Printing from moveable types.* This is the most important in its influence of all the arts of copying. It possesses a singular peculiarity, in the immense subdivision of the parts that form the pattern. After that pattern has furnished thousands of copies, the same individual elements may be arranged again and again in other forms, and thus supply multitudes of originals, from each of which thousands of their copied impressions may flow. It also possesses this advantage, that woodcuts may be used along with the letterpress, and impressions taken from both at the same operation.

94. *Printing from stereotype.* This mode of producing copies is very similar to the preceding. There are two modes by which stereotype plates are produced. In that most generally adopted a mould is taken in plaster from the moveable types, and in this the stereotype plate is cast. Another method has been employed in France: instead of composing the work in moveable type, it was set up in moveable copper matrices; each matrix being in fact a piece of copper of the same size as the type, and having the impression of the letter sunk into its surface, instead of projecting in relief. A stereotype plate may, it is

[a] Section number corrected.

evident, be obtained at once from this arrangement of matrices. The objection to the plan is the great expense of keeping so large a collection of matrices.

As the original composition does not readily admit of change, stereotype plates can only be applied with / advantage to cases where an extraordinary number of copies are demanded, or where the work consists of figures, and it is of great importance to ensure accuracy. Trifling alterations may, however, be made in it from time to time; and thus mathematical tables may, by the gradual extirpation of error, at last become perfect. This mode of producing copies possesses, in common with that by moveable types, the advantage of admitting the use of woodcuts; the copy of the woodcut in the stereotype plate being equally perfect with that of the moveable type. This union is of considerable importance, and cannot be accomplished with engravings on copper.[a]

95. *Lettering books.* The gilt letters on the backs of books are formed by placing a piece of gold leaf upon the leather, and pressing upon it brass letters previously heated: these cause the gold immediately under them to adhere to the leather, whilst the rest of the metal is easily brushed away. When a great number of copies of the same volume are to be lettered, it is found to be cheaper to have a brass pattern cut with the whole of the proper title: this is placed in a press, and being kept hot, the covers, each having a small bit of leaf-gold placed in the proper position, are successively brought under the brass, and stamped. The lettering at the back of the volume in the reader's hand was executed in this manner.

96. *Calico printing from blocks.* This is a mode of copying, by surface printing, from the ends of small pieces of copper wire, of various forms, fixed into a block of wood. They are all of one uniform height, about the eighth part of an inch above the surface of / the wood, and are arranged by the maker into any required pattern. If the block be placed upon a piece of fine woollen cloth, on which ink of any colour has been uniformly spread, the projecting copper wires receive a portion, which they give up when applied to the calico to be printed. By the former method of printing on calico, only one colour could be used; but by this plan, after the flower of a rose, for example, has

[a] The following appears in the section of additions, p. [396], in the original:
M. Genoux, a French printer, has invented a substance resembling pasteboard, which, by pressure upon moveable types, takes an impression from them, which may be used as a mould for stereotype pages.

been printed with one set of blocks, the leaves may be printed of another colour by a different set.

97. *Printing oilcloth.* After the canvas, which forms the basis of oilcloth, has been covered with paint of one uniform tint, the remainder of the processes which it passes through, are a series of copyings by surface printing, from patterns formed upon wooden blocks very similar to those employed by the calico printer. Each colour requiring a distinct set of blocks, those oilcloths with the greatest variety of colours are most expensive.

There are several other varieties of printing which we shall briefly notice as arts of copying; which, although not strictly surface printing, yet are more allied to it than that from copperplates.

98. *Letter copying.* In one of the modes of performing this process, a sheet of very thin paper is damped, and placed upon the writing to be copied. The two papers are then passed through a rolling press, and a portion of the ink from one paper is transferred to the other. The writing is, of course, reversed by this process; but the paper to which it is transferred being thin, the characters are seen through it on the other side, in their proper position. / Another common mode of copying letters is by placing a sheet of paper covered on both sides with a substance prepared from lamp-black, between a sheet of thin paper and the paper on which the letter to be despatched is to be written. If the upper or thin sheet be written upon with any hard pointed substance, the word written with this style will be impressed from the black paper upon both those adjoining it. The translucency of the upper sheet, which is retained by the writer, is in this instance necessary to render legible the writing which is on the back of the paper. Both these arts are very limited in their extent, the former affording two or three, the latter from two to perhaps ten or fifteen copies at the same time.

99. *Printing on china.* This is an art of copying which is carried to a very great extent. As the surfaces to which the impression is to be conveyed are often curved, and sometimes even fluted, the ink, or paint, is first transferred from the copper to some flexible substance, such as paper, or an elastic compound of glue and treacle. It is almost immediately conveyed from this to the unbaked biscuit, to which it more readily adheres.

100. *Lithographic printing.* This is another mode of producing copies in almost unlimited number. The original which supplies the copies is a drawing made on a stone of a slightly porous nature; the ink

employed for tracing it is made of such greasy materials that when water is poured over the stone it shall not wet the lines of the drawing. When a roller covered with printing ink, which is of an oily nature, is passed over the stone previously wetted, the water prevents / this ink from adhering to the uncovered portions; whilst the ink used in the drawing is of such a nature that the printing ink adheres to it. In this state, if a sheet of paper be placed upon the stone, and then passed under a press, the printing ink will be transferred to the paper, leaving the ink used in the drawing still adhering to the stone.

101. There is one application of lithographic printing which does not appear to have received sufficient attention, and perhaps further experiments are necessary to bring it to perfection. It is the reprinting of works which have just arrived from other countries. A few years ago one of the Paris newspapers was reprinted at Brussels as soon as it arrived by means of lithography. Whilst the ink is yet fresh, this may easily be accomplished: it is only necessary to place one copy of the newspaper on a lithographic stone; and by means of great pressure applied to it in a rolling press, a sufficient quantity of the printing ink will be transferred to the stone. By similar means, the other side of the newspaper may be copied on another stone, and these stones will then furnish impressions in the usual way. If printing from stone could be reduced to the same price per thousand as that from moveable types, this process might be adopted with great advantage for the supply of works for the use of distant countries possessing the same language. For a single copy might be printed off with *transfer ink*, and thus an English work, for example, might be published in America from stone, whilst the original, printed from moveable types, made its appearance on the same day in England.

102. It is much to be wished that such a method / were applicable to the reprinting of facsimilies of old and scarce books. This, however, would require the sacrifice of two copies, since a leaf must be destroyed for each page. Such a method of reproducing a small impression of an old work, is peculiarly applicable to mathematical tables, the setting up of which in type is always expensive and liable to error: but how long ink will continue to be transferable to stone, from paper on which it has been printed, must be determined by experiment. The destruction of the greasy or oily portion of the ink in the character of old books, seems to present the greatest impediment; if one constituent only of the ink were removed by time, it might perhaps be hoped, that chemical means would ultimately be

discovered for restoring it: but if this be unsuccessful, an attempt might be made to discover some substance having a strong affinity for the carbon of the ink which remains on the paper, and very little for the paper itself.*

103. Lithographic prints have occasionally been executed in colours. In such instances a separate stone seems to have been required for each colour, and considerable care, or very good mechanism, must have been employed to adjust the paper to each stone. If any two kinds of ink should be discovered mutually inadhesive, one stone might be employed for two inks; or if the inking-roller for the second and subsequent colours had portions cut away corresponding to those parts of the stone inked by the previous ones, then several colours might be printed from the same stone: but these principles do not / appear to promise much, except for coarse subjects.

104. *Register printing.* It is sometimes thought necessary to print from a wooden block, or stereotype plate, the same pattern reversed upon the opposite side of the paper. The effect of this, which is technically called *Register printing*, is to make it appear as if the ink had penetrated through the paper, and rendered the pattern visible on the other side. If the subject chosen contains many fine lines, it seems at first sight extremely difficult to effect so exact a super position of the two patterns, on opposite sides of the same piece of paper, that it shall be impossible to detect the slightest deviation; yet the process is extremely simple. The block which gives the impression is always accurately brought down to the same place by means of a hinge; this spot is covered by a piece of thin leather stretched over it; the block is now inked, and being brought down to its place, gives an impression of the pattern to the leather: it is then turned back; and being inked a second time, the paper intended to be printed is placed upon the leather, when the block again descending, the upper surface of the paper is printed from the block, and its undersurface takes up the impression from the leather. It is evident that the perfection of this mode of printing depends in a great measure on finding some soft substance like leather, which will take as much ink as it ought from the block, and which will give it up most completely to paper. Impressions thus obtained are usually fainter on the lower side; and in order in some

* I possess a lithographic reprint of a page of a table, which appears, from the form of the type, to have been several years old.

measure to remedy this defect, rather more ink is put on the block at the first than at the second impression. /

Of copying by casting

105. The art of casting, by pouring substances in a fluid state into a mould which retains them until they become solid, is essentially an art of copying; the form of the thing produced depending entirely upon that of the pattern from which it was formed.

106. *Of casting iron and other metals.* Patterns of wood or metal made from drawings are the originals from which the moulds for casting are made: so that, in fact, the casting itself is a copy of the mould; and the mould is a copy of the pattern. In castings of iron and metals for the coarser purposes, and, if they are afterwards to be worked, even for the finer machines, the exact resemblance amongst the things produced, which takes place in many of the arts to which we have alluded, is not effected in the first instance, nor is this necessary. As the metals shrink in cooling, the pattern is made larger than the intended copy; and in extricating it from the sand in which it is moulded, some little difference will occur in the size of the cavity which it leaves. In smaller works, where accuracy is more requisite, and where few or no after operations are to be performed, a mould of metal is employed which has been formed with considerable care. Thus, in casting bullets, which ought to be perfectly spherical and smooth, an iron instrument is used, in which a cavity has been cut and carefully ground; and, in order to obviate the contraction in cooling, a *jet* is left which may supply the deficiency of metal arising from that cause, and which is afterwards cut off. The leaden toys for children are cast in brass moulds / which open, and in which have been graved or chiselled the figures intended to be produced.

107. A very beautiful mode of representing small branches of the most delicate vegetable productions in bronze has been employed by Mr Chantrey. A small strip of a fir-tree, a branch of holly, a curled leaf of broccoli, or any other vegetable production, is suspended by one end in a small cylinder of paper which is placed for support within a similarly formed tin case. The finest river silt, carefully separated from all the coarser particles, and mixed with water, so as to have the consistency of cream, is poured into the paper cylinder by small portions at a time, carefully shaking the plant a little after each

addition, in order that its leaves may be covered, and that no bubbles of air may be left. The plant and its mould are now allowed to dry, and the yielding nature of the paper allows the loamy coating to shrink from the outside. When this is dry it is surrounded by a coarser substance; and, finally, we have the twig with all its leaves embedded in a perfect mould. This mould is carefully dried, and then gradually heated to a red heat. At the ends of some of the leaves or shoots, wires have been left to afford airholes by their removal, and in this state of strong ignition a stream of air is directed into the hole formed by the end of the branch. The consequence is, that the wood and leaves which had been turned into charcoal by the fire, are now converted into carbonic acid by the current of air; and, after some time, the whole of the solid matter of which the plant consisted is completely removed, leaving a hollow mould, bearing on its interior all the minutest traces / of its late vegetable occupant. When this process is completed, the mould being still kept at nearly a red heat, receives the fluid metal, which, by its weight, either drives the very small quantity of air, which at that high temperature remains behind, out through the airholes, or compresses it into the pores of the very porous substance of which the mould is formed.

108. When the form of the object intended to be cast is such that the pattern cannot be extricated from its mould of sand or plaster, it becomes necessary to make the pattern with wax, or some other easily fusible substance. The sand or plaster is moulded round this pattern, and, by the application of heat, the wax is extricated through an opening left purposely for its escape.

109. It is often desirable to ascertain the form of the internal cavities, inhabited by molluscous animals, such as those of spiral shells, and of the various corals. This may be accomplished by filling them with fusible metal, and dissolving the substance of the shell by muriatic acid; thus a metallic solid will remain which exactly filled all the cavities. If such forms are required in silver, or any other difficulty fusible metal, the shells may be filled with wax or resin, then dissolved away; and the remaining waxen form may serve as the pattern from which a plaster mould may be made for casting the metal. Some nicety will be required in these operations; and perhaps the minuter cavities can only be filled under an exhausted receiver.

110. *Casting in plaster.* This is a mode of copying applied to a variety of purposes: to produce / accurate representations of the human form – of statues – or of rare fossils – to which latter purpose it has lately

been applied with great advantage. In all casting, the first process is to make the mould; and plaster is the substance which is almost always employed for the purpose. The property which it possesses of remaining for a short time in a state of fluidity, renders it admirably adapted to this object, and adhesion, even to an original of plaster, is effectually prevented by oiling the surface on which it is poured. The mould formed round the subject which is copied, removed in separate pieces and then reunited, is that in which the copy is cast. This process gives additional utility and value to the finest works of art. The students of the Academy at Venice are thus enabled to admire the sculptured figures of Egina, preserved in the gallery at Munich; as well as the marbles of the Parthenon, the pride of our own Museum. Casts in plaster of the Elgin marbles adorn many of the academies of the Continent; and the liberal employment of such presents affords us an inexpensive and permanent source of popularity.

111. *Casting in wax.* This mode of copying, aided by proper colouring, offers the most successful imitations of many objects of natural history, and gives an air of reality to them which might deceive even the most instructed. Numerous figures of remarkable persons, having the face and hands formed in wax, have been exhibited at various times; and the resemblances have, in some instances been most striking. But whoever would see the art of copying in wax carried to the highest perfection, / should examine the beautiful collection of fruit at the house of the Horticultural Society; the model of the magnificent flower of the new genus Rafflesia – the waxen models of the internal parts of the human body which adorn the anatomical gallery of the Jardin des Plantes at Paris, and the Museum at Florence – or the collection of morbid anatomy at the University of Bologna. The art of imitation by wax does not usually afford the multitude of copies which flow from many similar operations. This number is checked by the subsequent stages of the process, which, ceasing to have the character of copying by a tool or pattern, become consequently more expensive. In each individual production, form alone is given by casting; the colouring must be the work of the pencil, guided by the skill of the artist.

Of copying by moulding

112. This method of producing multitudes of individuals having an exact resemblance to each other in external shape, is adopted very widely in the arts. The substances employed are, either naturally or by artificial preparation, in a soft or plastic state; they are then compressed by mechanical force, sometimes assisted by heat, into a mould of the required form.

113. *Of bricks and tiles.* An oblong box of wood fitting upon a bottom fixed to the brickmaker's bench, is the mould from which every brick is formed. A portion of the plastic mixture of which the bricks consist is made ready by less skilful hands: the workman first sprinkles a little sand / into the mould, and then throws the clay into it with some force; at the same time rapidly working it with his fingers, so as to make it completely close up to the corners. He next scrapes off, with a wetted stick, the superfluous clay, and shakes the new-formed brick dexterously out of its mould upon a piece of board, on which it is removed by another workman to the place appointed for drying it. A very skilful moulder has occasionally, in a long summer's day, delivered from ten to eleven thousand bricks; but a fair average day's work is from five to six thousand. Tiles of various kinds and forms are made of finer materials, but by the same system of moulding. Among the ruins of the city of Gour, the ancient capital of Bengal, bricks are found having projecting ornaments in high relief: these appear to have been formed in a mould, and subsequently glazed with a coloured glaze. In Germany, also, brickwork has been executed with various ornaments. The cornice of the church of St Stephano, at Berlin, is made of large blocks of brick moulded into the form required by the architect. At the establishment of Messrs Cubitt, in Gray's Inn Lane, vases, cornices, and highly ornamented capitals of columns are thus formed which rival stone itself in elasticity, hardness, and durability.[a]

114. *Of embossed china.* Many of the forms given to those beautiful

[a] The following appears in the section of additions, p. [396], in the original:
More recently a great variety of architectural ornaments, rosettes, arabesques, etc., have been cast with an alloy of zinc and tin. The foundry of Mr Geiss, at Berlin, supplies such ornaments in great perfection, and many new buildings in that capital are already ornamented with them.

specimens of earthenware which constitute the equipage of our breakfast and our dinner-tables, cannot be executed in the lathe of the potter. The embossed ornaments on the edges of the plates, their polygonal shape, the fluted surface of many of the vases, would all be difficult and / costly of execution by the hand; but they become easy and comparatively cheap, when made by pressing the soft material out of which they are formed into a hard mould. The care and skill bestowed on the preparation of that mould are repaid by the multitude it produces. In many of the works of the china manufactory, one part only of the article is moulded; the upper surface of the plate, for example, whilst the under side is figured by the lathe. In some instances, the handle, or only a few ornaments, are moulded, and the body of the work is turned.

115. *Glass seals.* The process of engraving upon gems requires considerable time and skill. The seals thus produced can therefore never become common. Imitations, however, have been made of various degrees of resemblance. The colour which is given to glass is, perhaps, the most successful part of the imitation. A small cylindrical rod of coloured glass is heated in the flame of a blowpipe, until the extremity becomes soft. The operator then pinches it between the ends of a pair of nippers, which are formed of brass, and on one side of which the device intended for the seal has been carved in relief. When the mould has been well finished and care is taken in heating the glass properly, the seals thus produced are not bad imitations; and by this system of copying they are so multiplied, that the more ordinary kinds are sold at Birmingham for three pence a dozen.

116. *Square glass bottles.* The round forms which are usually given to vessels of glass are readily produced by the expansion of the air with which they are blown. It is, however, necessary in / many cases to make bottles of a square form, and each capable of holding exactly the same quantity of fluid. It is also frequently desirable to have imprinted on them the name of the maker of the medicine or other liquid they are destined to contain. A mould of iron, or of copper, is provided of the intended size, on the inside of which are engraved the names required. This mould, which is used in a hot state, opens into two parts, to allow the insertion of the round, unfinished bottle, which is placed in it in a very soft state before it is removed from the end of the iron tube with which it was blown. The mould is now closed, and the glass is forced against its sides, by blowing strongly into the bottle.

117. *Wooden snuff boxes.* Snuff boxes ornamented with devices, in imitation of carved work or of rose engine turning, are sold at a price which proves that they are only imitations. The wood, or horn, out of which they are formed, is softened by long boiling in water, and whilst in this state it is forced into moulds of iron, or steel, on which are cut the requisite patterns, where it remains exposed to great pressure until it is dry.

118. *Horn knife handles and umbrella handles.* The property which horn possesses of becoming soft by the action of water and of heat, fits it for many useful purposes. It is pressed into moulds, and becomes embossed with figures in relief, adapted to the objects to which it is to be applied. If curved, it may be straightened; or if straight, it may be bent into any forms which ornament or utility may require; and by the use of the mould these forms may be / multiplied in endless variety. The commoner sorts of knives, the crooked handles for umbrellas, and a multitude of other articles to which horn is applied, attest the cheapness which the art of copying gives to the things formed of this material.

119. *Moulding tortoise-shell.* The same principle is applied to things formed out of the shell of the turtle, or the land tortoise. From the greatly superior price of the raw material, this principle of copying is, however, more rarely employed upon it; and the few carvings which are demanded, are usually performed by hand.

120. *Tobacco-pipe making.* This simple art is almost entirely one of copying. The moulds are formed of iron, in two parts, each embracing one half of the stem; the line of junction of these parts may generally be observed running lengthwise from one end of the pipe to the other. The hole passing to the bowl is formed by thrusting a long wire through the clay before it is enclosed in the mould. Some of the moulds have figures, or names, sunk in the inside, which give a corresponding figure in relief upon the finished pipe.

121. *Embossing upon calico.* Calicoes of one colour, but embossed all over with raised patterns, though not much worn in this country, are in great demand in several foreign markets. This appearance is produced by passing them between rollers, on one of which is figured in intaglio the pattern to be transferred to the calico. The substance of the cloth is pressed very forcibly into the cavities thus formed, and retains its pattern after considerable use. The *watered* appearance in the cover of the volume in the / reader's hands is produced in a similar manner. A cylinder of gun-metal, on which the design of the *watering*

is previously cut, is pressed by screws against another cylinder, formed out of pieces of brown paper which have been strongly compressed together and accurately turned. The two cylinders are made to revolve rapidly, the paper one being slightly damped, and, after a few minutes, it takes an impression from the upper or metal one. The glazed calico is now passed between the rollers, its glossy surface being in contact with the metal cylinder, which is kept hot by a heated iron enclosed within it. Calicoes are sometimes *watered* by placing two pieces on each other in such a position that the longitudinal threads of the one are at right angles to those of the other, and compressing them in this state between flat rollers. The threads of the one piece produce indentations in those of the other, but they are not so deep as when produced by the former method.

122. *Embossing upon leather.* This art of copying from patterns previously engraved on steel rollers is in most respects similar to the preceding. The leather is forced into the cavities, and the parts which are not opposite to any cavity are powerfully condensed between the rollers.

123. *Swaging.* This is an art of copying practised by the smith. In order to fashion his iron and steel into the various forms demanded by his customers, he has small blocks of steel into which are sunk cavities of different shapes; these are called *swages*, and are generally in pairs. Thus if he wants a round bolt, terminating in a cylindrical head of / larger diameter, and having one or more projecting rims, he uses a corresponding *swaging tool*; and having heated the end of his iron rod, and thickened it by striking the end in the direction of the axis (which is technically called *upsetting*), he places its head upon one part of the *swage*; and whilst an assistant holds the other part on the top of the hot iron, he strikes it several times with his hammer, occasionally turning the head one quarter round. The heated iron is thus forced by the blows to assume the form of the mould into which it is impressed.

124. *Engraving by pressure.* This is one of the most beautiful examples of the art of copying carried to an almost unlimited extent; and the delicacy with which it can be executed, and the precision with which the finest traces of the graving tool can be transferred from steel to copper, or even from hard steel to soft steel, is most unexpected. We are indebted to Mr Perkins for most of the contrivances which have brought this art at once almost to perfection. An engraving is first made upon soft steel, which is hardened by a

peculiar process without in the least injuring its delicacy. A cylinder of soft steel, pressed with great force against the hardened steel engraving, is now made to roll very slowly backward and forward over it, thus receiving the design, but in relief. The cylinder is in its turn hardened without injury; and if it be slowly rolled to and fro with strong pressure on successive plates of copper, it will imprint on a thousand of them a perfect facsimile of the original steel engraving from which it was made. Thus the number of copies / producible from the same design may be multiplied a thousand-fold. But even this is very far short of the limits to which the process may be extended. The hardened steel roller, bearing the design upon it in relief may be employed to make a few of its first impressions upon plates of *soft steel*, and these being hardened become the representatives of the original engraving, and may in their turn be made the parents of other rollers, each generating copperplates like their prototype. The possible extent to which facsimiles of one original engraving may thus be multiplied, almost confounds the imagination, and appears to be for all practical purposes unlimited.

This beautiful art was first proposed by Mr Perkins for the purpose of rendering the forgery of bank notes a matter of great difficulty; and there are two principles which peculiarly adapt it to that object: first, the perfect identity of all the impressions, so that any variation in the minutest line would at once cause detection; secondly, that the original plates may be formed by the united labours of several artists most eminent in their respective departments; for as only one original of each design is necessary, the expense, even of the most elaborate engraving, will be trifling, compared with the multitude of copies produced from it.

125. It must, however, be admitted that the principle of copying itself furnishes an expedient for imitating any engraving or printed pattern, however complicated; and thus presents a difficulty which none of the schemes devised for the prevention of forgery appear to have yet effectually obviated. In attempting to imitate the most perfect bank-note, the / first process would be to place it with the printed side downwards upon a stone or other substance, on which, by passing it through a rolling-press, it might be firmly fixed. The next object would be to discover some solvent which should dissolve the paper, but neither affect the printing-ink, nor injure the stone or substance to which it is attached. Water does not seem to do this effectually, and perhaps weak alkaline or acid solutions would be tried. If,

however, this could be fully accomplished, and if the stone or other substance, used to retain the impression, had those properties which enable us to print from it, innumerable facsimiles of the note might obviously be made, and the imitation would be complete. Porcelain biscuit, which has recently been used with a black lead pencil for memorandum books, seems in some measure adapted for such trials, since its porosity may be diminished to any required extent by regulating the dilution of the glazing.

126. *Gold and silver moulding.* Many of the mouldings used by jewellers consist of thin slips of metal, which have received their form by passing between steel rollers, on which the pattern is embossed or engraved; thus taking a succession of copies of the devices intended.

127. *Ornamental papers.* Sheets of paper coloured or covered with gold or silver leaf, and embossed with various patterns, are used for covering books, and for many ornamental purposes. The figures upon these are produced by the same process, that of passing the sheets of paper between engraved rollers. /

Of copying by stamping

128. This mode of copying is extensively employed in the arts. It is generally executed by means of large presses worked with a screw and heavy flywheel. The materials on which the copies are impressed are most frequently metals, and the process is sometimes executed when they are hot, and in one case when the metal is in a state between solidity and fluidity.

129. *Coins and medals.* The whole of the coins which circulate as money are produced by this mode of copying. The screw presses are either worked by manual labour, by water, or by steam power. The mint which was sent a few years since to Calcutta was capable of coining 200,000 pieces a day. Medals, which usually have their figures in higher relief than coins, are produced by similar means; but a single blow is rarely sufficient to bring them to perfection, and the compression of the metal which arises from the first blow renders it too hard to receive many subsequent blows without injury to the die. It is therefore, after being struck, removed to a furnace, in which it is carefully heated red-hot and annealed, after which operation it is again placed between the dies, and receives additional blows. For medals, on which the figures are very prominent, these processes must be

repeated many times. One of the largest medals hitherto struck underwent them nearly a hundred times before it was completed.

130. *Ornaments for military accoutrements, and furniture.* These are usually of brass, and are stamped up out of solid or sheet brass by placing it / between dies, and allowing a heavy weight to drop upon the upper die from a height of from five to fifteen feet.

131. *Buttons and nail heads.* Buttons embossed with crests or other devices are produced by the same means; and some of those which are plain receive their hemispherical form from the dies in which they are struck. The heads of several kinds of nails which are portions of spheres, or polyhedrons, are also formed by these means.

132. *Of a process for copying, called in France clichée.* This curious method of copying by stamping is applied to medals, and in some cases to forming stereotype plates. There exists a range of temperature previous to the melting point of several of the alloys of lead, tin, and antimony, in which the compound is neither solid, nor yet fluid. In this kind of pasty state it is placed in a box under a die, which descends upon it with considerable force. The blow drives the metal into the finest lines of the die, and the coldness of the latter immediately solidifies the whole mass. A quantity of the half-melted metal is scattered in all directions by the blow, and is retained by the sides of the box in which the process is carried on. The work thus produced is admirable for its sharpness, but has not the finished form of a piece just leaving the coining-press: the sides are ragged, and it must be trimmed, and its thickness equalized in the lathe.[a]

Of copying by punching

133. This mode of copying consists in driving a steel punch through the substance to be cut, either / by a blow or by pressure. In some

---

[a] The following appears in the section of additions, p. [397], in the original:

The alloy which is used for this purpose melts at the boiling point of water, or even below it.

Various degrees of fusibility are requisite for different purposes. The most usual proportions are:

	Bismuth	*Lead*	*Tin*	*Melting point in Fahrenheit*
For stamping ornaments	3	2	1	201
For cameos, etc.	8	5	3	190

If the metal is required to be sonorous, 5 oz. tin, 8 oz. bismuth, 8 oz. zinc, 8 oz. brass, and 8 of nitre, are taken. This alloy melts at about the heat of boiling water.

cases the object is to copy the aperture, and the substance separated from the plate is rejected; in other cases the small pieces cut out are the objects of the workman's labour.

134. *Punching iron plate for boilers.* The steel punch used for this purpose is from three-eighths to three-quarters of an inch in diameter, and drives out a circular disk from a plate of iron from one-quarter to five eighths of an inch thick.

135. *Punching tinned iron.* The ornamental patterns of open work which decorate the tinned and japanned wares in general use, are rarely punched by the workman who makes them. In London the art of punching out these patterns in screw-presses is carried on as a separate trade; and large quantities of sheet tin are thus perforated for cullenders, wine-strainers, borders of waiters, and other similar purposes. The perfection and the precision to which the art has been carried are remarkable. Sheets of copper, too, are punched with small holes about the hundredth of an inch in diameter, in such multitudes that more of the sheet metal is removed than remains behind; and plates of tin have been perforated with above three thousand holes in each square inch.

136. The inlaid plates of brass and rosewood, called *buhl work*, which ornament our furniture, are, in some instances, formed by punching; but in this case, both the parts cut out, and those which remain, are in many cases employed. In the remaining illustrations of the art of copying by punching, the part made use of is that which is punched out.

137. *Cards for guns.* The substitution of a / circular disk of thin card instead of paper, for retaining in its place the charge of a fowling-piece, is attended with considerable advantage. It would, however, be of little avail, unless an easy method was contrived of producing an unlimited number of cards, each exactly fitting the bore of the barrel. The small steel tool used for this purpose cuts out innumerable circles similar to its cutting end, each of which precisely fills the barrel for which it was designed.

138. *Ornaments of gilt paper.* The golden stars, leaves, and other devices, sold in shops for the purpose of ornamenting articles made of paper and pasteboard, and other fancy works, are cut by punches of various forms out of sheets of gilt paper.

139. *Steel chains.* The chain used in connecting the mainspring and fusee in watches and clocks, is composed of small pieces of sheet steel, and it is of great importance that each of these pieces should be

of exactly the same size. The links are of two sorts; one of them consisting of a single oblong piece of steel with two holes in it, and the other formed by connecting two of the same pieces of steel, placed parallel to each other, and at a small distance apart, by two rivets. The two kinds of links occur alternately in the chain: each end of the single pieces being placed between the ends of two others, and connected with them by a rivet passing through all three. If the rivet holes in the pieces for the double links are not precisely at equal distances, the chain will not be straight, and will, consequently, be unfit for its purpose. /

Copying with elongation

140. In this species of copying there exists but little resemblance between the copy and the original. It is the cross-section only of the thing produced which is similar to the tool through which it passes. When the substances to be operated upon are hard, they must frequently pass in succession through several holes, and it is in some cases necessary to anneal them at intervals.

141. *Wire drawing*. The metal to be converted into wire is made of a cylindrical form, and drawn forcibly through circular holes in plates of steel: at each passage it becomes smaller; and, when finished, its section at any point is a precise copy of the last hole through which it passed. Upon the larger kinds of wire, fine lines may sometimes be traced, running longitudinally; these arise from slight imperfections in the holes of the draw-plates. For many purposes of the arts, wire, the section of which is square or half round, is required: the same method of making it is pursued, except that the holes through which it is drawn are in such cases themselves square, or half-round, or of whatever other form the wire is required to be. A species of wire is made, the section of which resembles a star with from six to twelve rays; this is called pinion wire, and is used by the clockmakers. They file away all the rays from a short piece, except from about half an inch near one end: this becomes a pinion for a clock; and the leaves or teeth are already burnished and finished, from having passed through the *draw-plate*.

142. *Tube drawing*. The art of forming tubes / of uniform diameter is nearly similar in its mode of execution to wire drawing. The sheet brass is bent round and soldered so as to form a hollow cylinder; and

if the diameter outside is that which is required to be uniform, it is drawn through a succession of holes, as in wire drawing. If the inside diameter is to be uniform, a succession of steel cylinders, called *triblets*, are drawn through the brass tube. In making tubes for telescopes, it is necessary that both the inside and outside should be uniform. A steel *triblet*, therefore, is first passed into the tube, which is then drawn through a succession of holes, until the outside diameter is reduced to the required size. The metal of which the tube is formed is condensed between these holes and the steel cylinder within; and when the latter is withdrawn the internal surface appears polished. The brass tube is considerably extended by this process, sometimes even to double its first length.

143. *Leaden pipes.* Leaden pipes for the conveyance of water were formerly made by casting; but it has been found that they can be made both cheaper and better by drawing them through holes in the manner last described. A cylinder of lead, of five or six inches in diameter and about two feet long, is cast with a small hole through its axis, and an iron *triblet* of about fifteen feet in length is forced into the hole. It is then drawn through a series of holes, until the lead is extended upon the *triblet* from one end to the other, and is of the proper thickness in proportion to the size of the pipe.

144. *Iron rolling.* When cylinders of iron of greater thickness than wire are required, they are formed by passing wrought iron between rollers, each / of which has sunk in it a semicylindrical groove; and as such rollers rarely touch accurately, a longitudinal line will usually be observed in the cylinders so manufactured. Bar iron is thus shaped into all the various forms of round, square, half-round, oval, etc. in which it occurs in commerce. A particular species of moulding is thus made, which resembles, in its section, that part of the frame of a window which separates two adjacent panes of glass. Being much stronger than wood, it can be considerably reduced in thickness, and consequently offers less obstruction to the light; it is much used for skylights.

145. It is sometimes required that the iron thus produced should not be of uniform thickness throughout. This is the case in bars for railroads, where greater depth is required towards the middle of the rail which is at the greatest distance from the supports. This form is produced by cutting the groove in the rollers deeper at those parts where additional strength is required, so that the hollow which surrounds the roller would, if it could be unwound, be a mould of the shape the iron is intended to fit.

146. *Vermicelli.* The various forms into which this paste is made are given by forcing it through holes in tin plate. It passes through them, and appears on the other side in long strings. The cook makes use of the same method in preparing butter and ornamental pastry for the table, and the confectioner in forming cylindrical lozenges of various composition.

Of copying with altered dimensions

147. *Of the pentagraph.* This mode of copying is chiefly used for drawings or maps: the instrument is simple; and, although usually employed in reducing, / is capable of enlarging the size of the copy. An automaton figure, exhibited in London a short time since, which drew profiles of its visitors, was regulated by a mechanism on this principle. A small aperture in the wall, opposite the seat in which the person is placed whose profile is taken, conceals a camera lucida, which is placed in an adjoining apartment: and an assistant, by moving a point, connected by a pentagraph with the hand of the automaton, over the outline of the head, causes the figure to trace a corresponding profile.

148. *By turning.* The art of turning might perhaps itself be classed amongst the arts of copying. A steel axis, called a *mandril*, having a pulley attached to the middle of it, is supported at one end either by a conical point, or by a cylindrical collar, and at the other end by another *collar*, through which it passes. The extremity which projects beyond this last *collar* is formed into a screw, by which various instruments, called *chucks*, can be attached to it. These *chucks* are intended to hold the various materials to be submitted to the operation of turning, and have a great variety of forms. The *mandril* with the *chuck* is made to revolve by a strap which passes over the pulley that is attached to it, and likewise over a larger wheel moved either by the foot, or by its connection with steam or water power. All work which is executed on a *mandril* partakes in some measure of the irregularities in the form of that *mandril*; and the perfect circularity of section which ought to exist in every part of the work, can only be ensured by an equal accuracy in the *mandril* and its *collar*.

149. *Rose engine turning.* This elegant art depends / in a great measure on copying. Circular plates of metal called *rosettes*, having various indentations on the surfaces and edges, are fixed on the

mandril, which admits of a movement either end-wise or laterally: a fixed obstacle called the '*touch*', against which the rosettes are pressed by a spring, obliges the mandril to follow their indentations, and thus causes the cutting tool to trace out the same pattern on the work. The distance of the cutting tool from the centre being usually less than the radius of the *rosette*, causes the copy to be much diminished.

150. *Copying dies.* A lathe has been long known in France, and recently been used at the English mint for copying dies. A blunt point is carried by a very slow spiral movement successively over every part of the die to be copied, and is pressed by a weight into all the cavities; while a cutting point connected with it by the machine traverses the face of a piece of soft steel, in which it cuts the device of the original die on the same or on a diminished scale. The degree of excellence of the copy increases in proportion as it is smaller than the original. The die of a crown-piece will furnish by copy a very tolerable die for a sixpence. But the chief use to be expected from this lathe is to prepare all the coarser parts, and leave only the finer and more expressive lines for the skill and genius of the artist.

151. *Shoe-last making engine.* An instrument not very unlike in principle was proposed for the purpose of making shoe lasts. A pattern last of a shoe for the right foot was placed in one part of the apparatus, and when the machine was moved, two pieces of wood, placed in another part which had been previously / adjusted by screws, were cut into lasts greater or less than the original, as was desired; and although the pattern was for the right foot, one of the lasts was for the left, an effect which was produced by merely interposing a wheel which reversed the motion between the two pieces of wood to be cut into lasts.

152. *Engine for copying busts.* Many years since, the late Mr Watt amused himself with constructing an engine to produce copies of busts or statues, either of the same size as the original, or in a diminished proportion. The substances on which he operated were various, and some of the results were shown to his friends, but the mechanism by which they were made has never been described. More recently, Mr Hawkins, who, nearly at the same time, had also contrived a similar machine, has placed it in the hands of an artist, who has made copies in ivory from a variety of busts. The art of multiplying in *different sizes* the figures of the sculptor, aided by that of rendering their acquisition cheap through the art of casting,

promises to give additional value to his productions, and to diffuse more widely the pleasure arising from their possession.

153. *Screw cutting.* When this operation is performed in the lathe by means of a screw upon the *mandril*, it is essentially an art of copying, but it is only the number of threads in a given length which is copied; the *form* of the thread, and length as well as the diameter of the screw to be cut, are entirely independent of those from which the copy is made. There is another method of cutting screws in a lathe by means of one pattern screw, which, being connected by wheels with the *mandril*, guides the cutting point. / In this process, unless the time of revolution of the *mandril* is the same as that of the screw which guides the cutting point, the number of threads in a given length will be different. If the *mandril* move quicker than the cutting point, the screw which is produced will be finer than the original; if it move slower, the copy will be more coarse than the original. The screw thus generated may be finer or coarser – it may be larger or smaller in diameter – it may have the same or a greater number of threads than that from which it is copied; yet all the defects which exist in the original will be accurately transmitted, under the modified circumstances, to every individual generated from it.

154. *Printing from copper plates with altered dimensions.* Some very singular specimens of an art of copying, not yet made public, were brought from Paris a few years since. A watchmaker in that city, of the name of Gonord, had contrived a method by which he could take from the same copperplate impressions of different sizes, either larger or smaller than the original design. Having procured four impressions of a parrot, surrounded by a circle, executed in this manner, I showed them to the late Mr Lowry, an engraver equally distinguished for his skill, and for the many mechanical contrivances with which he enriched his art. The relative dimensions of the several impressions were 5·5, 6·3, 8·4, 15·0, so that the largest was nearly three times the linear size of the smallest; and Mr Lowry assured me, that he was unable to detect any lines in one which had not corresponding lines in the others. There appeared to be a difference in the quantity of ink, but none in / the traces of the engraving; and, from the general appearance, it was conjectured that the largest but one was the original impression from the copperplate.

The means by which this singular operation was executed have not been published; but two conjectures were formed at the time which merit notice. It was supposed that the artist was in possession of some method of transferring the ink from the lines of a copperplate to the surface of some fluid, and of retransferring the impression from the fluid to paper. If this could be accomplished, the print would, in the first instance, be of exactly the same size as the copper from which it was derived; but if the fluid were contained in a vessel having the form of an inverted cone, with a small aperture at the bottom, the liquid might be lowered or raised in the vessel by gradual abstraction or addition through the apex of the cone; in this case, the surface to which the printing-ink adhered would diminish or enlarge, and in this altered state the impression might be retransferred to paper. It must be admitted, that this conjectural explanation is liable to very considerable difficulties; for, although the converse operation of taking an impression from a liquid surface has a parallel in the art of marbling paper, the possibility of transferring the ink from the copper to the fluid requires to be proved.

Another and more plausible explanation is founded on the elastic nature of the compound of glue and treacle, a substance already in use in transferring engravings to earthenware. It is conjectured, that an impression from the copperplate is taken upon a large sheet of this composition; that this sheet is / then stretched in both directions, and that the ink thus expanded is transferred to paper. If the copy is required to be smaller than the original, the elastic substance must first be stretched, and then receive the impression from the copperplate: on removing the tension it will contract, and thus reduce the size of the design. It is possible that one transfer may not in all cases suffice; as the extensibility of the composition of glue and treacle, although considerable, is still limited. Perhaps sheets of India rubber of uniform texture and thickness, may be found to answer better than this composition; or possibly the ink might be transferred from the copper plate to the surface of a bottle of this gum, which bottle might, after being expanded by forcing air into it, give up the enlarged impression to paper. As it would require considerable time to produce impressions in this manner, and there might arise some difficulty in making them all of precisely the same size, the process might be rendered more certain and expeditious by performing that part of the operation which depends on the enlargement or diminution of the design only once; and, instead of printing from the soft substance, transferring the design from it to stone: thus a considerable portion of the work would be reduced to an art already well known, that of lithography. This idea receives some confirmation from the fact, that in another set of specimens, consisting of a map of St Petersburgh, of several sizes, a very short line, evidently an accidental defect, occurs in all the impressions of one particular size, but not in any of a different size.

155. *Machine to produce engraving from medals.* / An instrument was contrived, a long time ago, and is described in the *Manuel de Tourneur*, by which copperplate engravings are produced from medals and other objects in relief. The medal and the copper are fixed on two sliding plates at right angles to each other, so connected that, when the plate on which the medal is fixed is raised vertically by a screw, the slide holding the copperplate is advanced by an equal quantity in the horizontal direction. The medal is fixed on the vertical slide with its face towards the copperplate, and a little above it.

A bar, terminating at one end in a tracing point, and at the other in a short arm, at right angles to the bar, and holding a diamond point, is placed horizontally above the copper; so that the tracing point shall touch the medal to which the bar is perpendicular, and the diamond point shall touch the copperplate to which the arm is perpendicular.

Under this arrangement, the bar being supposed to move parallel to itself, and consequently to the copper, if the tracing point pass over a flat part of the medal, the diamond point will draw a straight line of equal length upon the copper; but, if the tracing point pass over any projecting part of the medal, the deviation from the straight line by the diamond point, will be exactly equal to the elevation of the corresponding point of the medal above the rest of the surface. Thus, by the transit of this tracing point over any line upon the medal, the diamond will draw upon the copper a section of the medal through that line.

A screw is attached to the apparatus, so that if the medal be raised a very small quantity by the screw, / the copperplate will be advanced by the same quantity, and thus a new line of section may be drawn: and, by continuing this process, the series of sectional lines on the copper produces the representation of the medal on a plane: the outline and the form of the figure arising from the sinuosities of the lines, and from their greater or less proximity. The effect of this kind of engraving is very striking; and in some specimens gives a high degree of apparent relief. It has been practised on plate glass, and is then additionally curious from the circumstance of the fine lines traced by the diamond being invisible, except in certain lights.

From this description, it will have been seen that the engraving on copper must be distorted; that is to say, that the projection on the copper cannot be the same as that which arises from a perpendicular projection of each point of the medal upon a plane parallel to itself. The position of the prominent parts will be more altered than that of the less elevated; and the greater the relief of the medal the more distorted will be its engraved representation. Mr John Bate, son of Mr Bate, of the Poultry, has contrived an improved machine, for which he has taken a patent, in which this source of distortion is remedied. The head, in the title page of the present volume, is copied from a medal of Roger Bacon, which forms one of a series of medals of eminent men, struck at the Royal Mint at Munich, and is the first of the published productions of this new art.* /

The inconvenience which arises from too high a relief in the medal, or in the bust, might be remedied by some mechanical contrivance, by which the deviation of the diamond point from the right line (which it would describe when the tracing point traverses a plane), would be made proportional – not to the elevation of the corresponding point above the plane of the medal, but to its elevation above some other parallel plane removed to a fit distance behind it. Thus busts and statues might be reduced to any required degree of relief.

156. The machine just described naturally suggests other views which seem to deserve some consideration, and, perhaps, some experiment. If a medal were placed under the tracing point of a pentagraph, an engraving tool substituted for the pencil, and a copperplate in the place of the paper; and if, by some mechanism, the tracing point, which slides in a vertical plane, could, as it is carried over the different elevations of the medal, increase or diminish the depth of the engraved line proportionally to the actual height of the corresponding point on the medal, then an engraving would be produced, free at least from any distortion, although it might be liable to objections of a different kind. If, by any similar contrivance, instead of lines, we could make on each point of the copper a dot, varying in size or depth with the altitude of the corresponding point of the medal above its plane, than a new species of engraving would

* The construction of the engraving becomes evident on examining it with a lens of sufficient power to show the continuity of the lines.

be produced: and the variety of these might again be increased, by causing the graving point to describe very small circles, of diameters, varying with the height of the point / on the medal above a given plane; or by making the graving tool consist of three equidistant points, whose distance increased or diminished according to some determinate law, dependent on the elevation of the point represented above the plane of the medal. It would, perhaps, be difficult to imagine the effects of some of these kinds of engraving; but they would all possess, in common, the property of being projections, by parallel lines, of the objects represented, and the intensity of the shade of the ink would either vary according to some function of the distance of the point represented from some given plane, or it would be a little modified by the distances from the same plane of a few of the immediately contiguous points.

157. The system of shading maps by means of lines of equal altitude above the sea bears some analogy to this mode of representing medals, and if applied to them would produce a different species of engraved resemblance. The projections on the plane of the medal, of the section of an imaginary plane, placed at successive distances above it, with the medal itself, would produce a likeness of the figure on the medal, in which all the inclined parts of it would be dark in proportion to their inclination. Other species of engraving might be conceived by substituting, instead of the imaginary plane, an imaginary sphere or other solid, intersecting the figure in the medal.

158. *Lace made by caterpillars.* A most extraordinary species of manufacture, which is in a slight degree connected with copying, has been contrived by an officer of engineers residing at Munich. It consists of lace, and veils, with open patterns in / them, made entirely by caterpillars. The following is the mode of proceeding adopted: he makes a paste of the leaves of the plant, which is the usual food of the species of caterpillar* he employs, and spreads it thinly over a stone, or other flat substance. He then, with a camel-hair pencil dipped in olive oil, draws upon the coating of paste the pattern he wishes the insects to leave open. This stone is then placed in an inclined position, and a number of the caterpillars are placed at the bottom. A peculiar species is chosen, which spins a strong web; and the animals commencing at the bottom, eat and spin their way up to the top, carefully avoiding every part touched by the oil, but devouring all the

* The *Phalaena pardilla*, which feeds on the *Prunus padus*.

rest of the paste. The extreme lightness of these veils, combined with some strength, is truly surprising. One of them, measuring twenty-six and a half inches by seventeen inches, weighed only 1·51 grains; a degree of lightness which will appear more strongly by contrast with other fabrics. One square yard of the substance of which these veils are made weighs 4⅓ grains, whilst one square yard of silk gauze weighs 137 grains, and one square yard of the finest patent net weighs 262½ grains. The ladies' coloured muslin dresses, mentioned in the table subjoined, cost ten shillings per dress, and each weigh six ounces; the cotton from which they are made weighing nearly six and two-ninth ounces avoirdupois weight.[a] /

*Weight of one square yard of each of the following articles**

Description of goods	*Value per yard measure*	*Weight finished of one square yard*	*Weight of cotton used in making one square yard*
	s. d.	*Troy grains*	*Troy grains*
Caterpillar veils	—	4⅓	—
Silk gauze 3-4 wide	1 0	137	—
Finest patent net	—	262½	—
Fine cambric muslin	—	551	—
6-4ths jaconet muslin	2 0	613	670
Ladies' coloured muslin dresses	3 0	788	875
6-4ths cambric	1 2	972	1069
9-8ths calico	0 9	988	1085
½-yard nankeen	0 8	2240	2432

159. This enumeration, which is far from complete, of the arts in which copying is the foundation, may be terminated with an example which has long been under the eye of the reader; although few,

* Some of these weights and measures are calculated from a statement in the Report of the Committee of the House of Commons on Printed Cotton Goods; and the widths of the pieces there given are presumed to be the real widths, not those by which they are called in the retail shops.

[a] The following appears in the section of additions, p. [397], in the original: The caterpillar is said to be either the Tinea Punctata or Tinea Padilla. One insect can work ¼ inch of cloth, and the inventor of the art, Mr De Hebenstreit, in order to give more strength to the cloth, sometimes compels the caterpillars to work several times over the same piece.

perhaps, are aware of the number of repeated copyings of which these very pages are the subject.

1. They are copies, by printing, from stereotype plates.

2. These stereotype plates are copied, by the art of casting, from moulds formed of plaster of Paris. /

3. These moulds are themselves copied by casting the plaster in a liquid state upon the moveable types set up by the compositor.

[It is here that the union of the intellectual and the mechanical departments takes place. The mysteries, however, of an author's copying, form no part of our enquiry, although it may be fairly remarked, that, in numerous instances, the mental far eclipses the mechanical copyist.]

4. These moveable types, the obedient messengers of the most opposite thoughts, the most conflicting theories, are themselves copies by casting from moulds of copper called *matrices*.

5. The lower part of those *matrices*, bearing the impressions of the letters or characters, are copies, by punching, from steel punches on which the same characters exist in relief.

6. These steel punches are not themselves entirely exempted from the great principle of art. Many of the cavities which exist in them, such as those in the middle of the punches for the letters *a*, *b*, *d*, *e*, *g*, etc., are produced from other steel punches in which these parts are in relief.

We have thus traced through six successive stages of copying the mechanical art of printing from stereotype plates: the principle of copying contributing in this, as in every other department of manufacture, to the uniformity and the cheapness of the work produced./

CHAPTER XII

ON THE METHOD OF OBSERVING MANUFACTORIES

160. Having now reviewed the *mechanical* principles which regulate the successful application of mechanical science to great establishments for the production of manufactured goods, it remains for us to suggest a few enquiries, and to offer a few observations, to those whom an enlightened curiosity may lead to examine the factories of this or of other countries.

The remark – that *it is important to commit to writing all information as soon as possible after it is received, especially when numbers are concerned* – applies to almost all enquiries. It is frequently impossible to do this at the time of visiting an establishment, although not the slightest jealousy may exist; the mere act of writing information as it is communicated orally, is a great interruption to the examination of machinery. In such cases, therefore, it is advisable to have prepared beforehand the questions to be asked, and to leave blanks for the answers, which may be quickly inserted, as, in a multitude of cases, they are merely numbers. Those who have not tried this plan will be surprised at the quantity of information which may, through its means, be acquired, even by a short examination. Each / manufacture requires its own list of questions, which will be better drawn up after the first visit. The following outline, which is very generally applicable, may suffice for an illustration; and to save time, it may be convenient to have it printed; and to bind up, in the form of a pocket-book, a hundred copies of the skeleton forms for processes, with about twenty of the general enquiries.

GENERAL ENQUIRIES

Outlines of a description of any of the mechanical arts ought to contain information on the following points

Brief sketch of its history, particularly the date of its invention, and of its introduction into England.

Short reference to the previous states through which the material employed has passed; the places whence it is procured; the price of a given quantity.

[The various processes must now be described successively according to the plan which will be given in (161); after which the following information should be given.]

Are various kinds of the same article made in one establishment, or at different ones, and are there differences in the processes?

To what defects are the goods liable?

What substitutes or adulterations are used?

What waste is allowed by the master?

What tests are there of the goodness of the manufactured articles?

The weight of a given quantity, or number, and a comparison with that of the raw material?

The wholesale price at the manufactory? (£ *s.* *d.*) per ( )

The usual retail price? (£ *s.* *d.*)

Who provide tools? Master, or men? Who repair tools? Master, or men?

What is the expense of the machinery? /

What is the annual wear and tear, and what its duration?

Is there any particular trade for making it? Where?

Is it made and repaired at the manufactory?

In any manufactory visited, state the number ( ) of processes; and of the persons employed in each process; and the quantity of manufactured produce.

What quantity is made annually in Great Britain?

Is the capital invested in manufactories large or small?

Mention the principal seats of this manufacture in England; and if it flourishes abroad, the places where it is established.

The duty, excise, or bounty, if any, should be stated, and any alterations in past years; and also the amount exported or imported for a series of years.

Whether the same article, but of superior, equal, or inferior make, is imported?

Does the manufacturer export, or sell, to a middleman, who supplies the merchant?

To what countries is it chiefly sent? – and in what goods are the returns made?

161. Each process requires a separate skeleton, and the following outline will be sufficient for many different manufactories:

Process ( ) Manufacture ( )
Place ( ) Name ( )
date 183

The mode of executing it, with sketches of the tools or machine if necessary.
The number of persons necessary to attend the machine.
Are the operatives men, ( ) women, ( ) or children? ( ) If mixed, what are the proportions?
What is the pay of each? (*s.* *d.*) (*s.* *d.*) (*s.* *d.*) per ( )
What number ( ) of hours do they work per day? /
Is it usual, or necessary, to work night and day without stopping?
Is the labour performed by piece- or by day-work?
Who provide tools? Master, or men? Who repair tools? Master, or men?
What degree of skill is required, and how many years' ( ) apprenticeship?
The number of times ( ) the operation is repeated per day or per hour?
The number of failures ( ) in a thousand?
Whether the workmen or the master loses by the broken or damaged articles?
What is done with them?
If the same process is repeated several times, state the diminution or increase of measure, and the loss, if any, at each repetition.

162. In this skeleton, the answers to the questions are in some cases printed, as 'Who repair the tools? – *Masters*, *Men*'; in order that the proper answer may be underlined with a pencil. In filling up the answers which require numbers, some care should be taken: for instance, if the observer stands with his watch in his hand before a person heading a pin, the workman will almost certainly increase his speed, and the estimate will be too large. A much better average will result from enquiring what quantity is considered a fair day's work. When this cannot be ascertained, the number of operations performed in a given time may frequently be counted when the workman is quite unconscious that any person is observing him. Thus the sound made by the motion of a loom may enable the observer to count the number of strokes per minute, even though he is outside the building in which it is contained. / M. Coulomb, who had great experience in making such observations, cautions those who may repeat his experiments against being deceived by such circumstances: 'Je prie' (says he) 'ceux qui voudront les repeter, s'ils n'ont pas le temps de mesurer les resultats après plusiers jours d'un travail continu,

d'observer les ouvriers à différentes reprises dans la journée, sans qu'ils sachent qu'ils sont observés. L'on ne peut trop avertir combien l'on risque de se tromper en calculant, soit la vitesse, soit le temps effectif du travail, d'après une observation de quelques minutes.'[a] *Mémoires de l'Institut*. vol. II, p. 247. It frequently happens, that in a series of answers to such questions, there are some which, although given directly, may also be deduced by a short calculation from others that are given or known; and advantage should always be taken of these verifications, in order to confirm the accuracy of the statements; or, in case they are discordant, to correct the apparent anomalies. In putting lists of questions into the hands of a person undertaking to give information upon any subject, it is in some cases desirable to have an estimate of the soundness of his judgement. The questions can frequently be so shaped, that some of them may indirectly depend on others; and one or two may be inserted whose answers can be obtained by other methods: nor is this process without its advantages in enabling us to determine the value of our own judgement. The habit of forming an estimate of the magnitude of any object or the frequency of any occurrence, immediately previous to our applying to it measure or number, tends materially to fix the attention and to improve the judgement. /

[a] *Translation:* I ask those who would like to repeat them, should they not have the time to evaluate the results after several days of continuous work, to observe the workers at different intervals in the day, without them knowing that they are observed. We must warn very strongly that there is a great possibility of making errors in calculation, either of the speed, or of the effective period of work, if it is based on the observation of a few minutes.

SECTION II

On the domestic and political economy of manufactures

CHAPTER XIII

DISTINCTION BETWEEN MAKING AND MANUFACTURING

163. The *economical principles* which regulate the application of machinery, and which govern the interior of all our great factories, are quite as essential to the prosperity of a great commercial country, as are those mechanical principles, the operation of which has been illustrated in the preceding section.

The first object of every person who attempts to make any article of consumption, is, or ought to be, to produce it in a perfect form; but in order to secure to himself the greatest and most permanent profit, he must endeavour, by every means in his power, to render the new luxury or want which he has created, cheap to those who consume it. The larger number of purchasers thus obtained will, in some measure, / secure him from the caprices of fashion, whilst it furnishes a far greater amount of profit, although the contribution of each individual is diminished. The importance of collecting data, for the purpose of enabling the manufacturer to ascertain how many additional customers he will acquire by a given reduction in the price of the article he makes, cannot be too strongly pressed upon the attention of those who employ themselves in statistical enquiries. In some ranks of society, no diminution of price can bring forward a great additional number of customers; whilst, amongst other classes, a very small reduction will so enlarge the sale, as to yield a considerable increase of profit. Materials calculated to assist in forming a table of the numbers of persons who possess incomes of different amount, occur in the 14th Report of the Commissioners of Revenue Inquiry; which includes a statement of the amount of personal property proved at the legacy office during one year; the number of the various classes of testators; and an account of the number of persons receiving dividends from funded property, distributed into classes. Such a table, formed even approximately, and exhibited in the form of a curve, might be of service.

164. A considerable difference exists between the terms *making* and *manufacturing*. The former refers to the production of *a small*, the latter to that of *a very large number of individuals*; and the difference is well illustrated in the evidence, given before the Committee of the House of Commons, on the Export of Tools and Machinery. On that occasion Mr Maudslay stated, that he had been applied to by the Navy Board to make iron tanks for ships, and / that he was rather unwilling to do so, as he considered it to be out of his line of business; however, he undertook to make one as a trial. The holes for the rivets were punched by hand-punching with presses, and the 1680 holes which each tank required cost seven shillings. The Navy Board, who required a large number, proposed that he should supply forty tanks a week for many months. The magnitude of the order made it worth his while to commence *manufacture*, and to make tools for the express business. Mr Maudslay therefore offered, if the Board would give him an order for two thousand tanks, to supply them at the rate of eighty per week. The order was given: he made tools, by which the expense of punching the rivet-holes of each tank was reduced from seven shillings to ninepence; he supplied ninety-eight tanks a week for six months, and the price charged for each was reduced from seventeen pounds to fifteen.

165. If, therefore, the *maker* of an article wish to become a *manufacturer*, in the more extended sense of the term, he must attend to other principles besides those mechanical ones on which the successful execution of his work depends; and he must carefully arrange the whole system of his factory in such a manner, that the article he sells to the public may be produced at as small a cost as possible. Should he not be actuated at first by motives so remote, he will, in every highly civilized country, be compelled, by the powerful stimulus of competition, to attend to the principles of the domestic economy of manufactures. At every reduction in price of the commodity he makes, he will be driven to seek compensation / in a saving of expense in some of the processes; and his ingenuity will be sharpened in this enquiry by the hope of being able in his turn to undersell his rivals. The benefit of the improvements thus engendered is, for a short time, confined to those from whose ingenuity they derive their origin; but when a sufficient experience has proved their value, they become generally adopted, until in their turn they are superseded by other more economical methods. /

CHAPTER XIV

OF MONEY AS A MEDIUM OF EXCHANGE

166. In the earlier stages of societies the interchange of the few commodities required was conducted by barter; but as soon as their wants became more varied and extensive, the necessity of having some common measure of the value of all commodities – itself capable of subdivision – became apparent: thus money was introduced. In some countries shells have been employed for this purpose; but civilized nations have, by common consent, adopted the precious metals.* The sovereign power has, in most countries, assumed the right of coining; or, in other words, the right of stamping with distinguishing marks, pieces of metal having certain forms and weights and a certain degree of fineness: the marks becoming a guarantee, to the people amongst whom the money circulates, that each piece is of the required weight and quality. /

The expense of manufacturing gold into coin, and that of the loss arising from wear, as well as of interest on the capital invested in it, must either be defrayed by the State, or be compensated by a small reduction in its weight, and is a far less cost to the nation than the loss of time and inconvenience which would arise from a system of exchange or barter.

167. These coins are liable to two inconveniences: they may be manufactured privately by individuals, of the same quality, and similarly stamped; or imitations may be made of inferior metal, or of diminished weight. The first of these inconveniences would be easily remedied by making the current value of the coin nearly equal to that

* In Russia platinum has been employed for coin; and it possesses a peculiarity which deserves notice. Platinum cannot be melted in our furnaces, and is chiefly valuable in commerce when in the shape of ingots, from which it may be forged into useful forms. But when a piece of platinum is cut into two parts, it cannot easily be reunited except by means of a chemical process, in which both parts are dissolved in an acid. Hence when platinum coin is too abundant, it cannot, like gold, be reduced into masses by melting, but must pass through an expensive process to render it useful.

of the same weight of the metal; and the second would be obviated by the caution of individuals in examining the external characters of each coin, and partly by the punishment inflicted by the State on the perpetrators of such frauds.

168. The subdivisions of money vary in different countries, and much time may be lost by an inconvenient system of division. The effect is felt in keeping extensive accounts, and particularly in calculating the interest on loans, or the discount upon bills of exchange. The decimal system is the best adapted to facilitate all such calculations; and it becomes an interesting question to consider whether our own currency might not be converted into one decimally divided. The great step, that of abolishing the guinea, has already been taken without any inconvenience, and but little is now required to render the change complete.

169. If, whenever it becomes necessary to call / in the half-crowns, a new coin of the value of two shillings were issued, which should be called by some name implying a unit (a prince, for instance), we should have the tenth part of a sovereign. A few years after, when the public were familiar with this coin, it might be divided into one hundred instead of ninety-six farthings; and it would then consist of twenty-five pence, each of which would be four per cent. less in value than the former penny. The shillings and sixpences being then withdrawn from circulation, their place might be supplied with silver coins each worth five of the new pence, and by others of tenpence, and of twopence halfpenny; the latter coin, having a distinct name, would be the tenth part of a prince.

170. The various manufactured commodities, and the various property possessed by the inhabitants of a country, all become measured by the standard thus introduced. But it must be observed that the value of gold is itself variable; and that, like all other commodities, its price depends on the extent of the demand compared with that of the supply.

171. As transactions multiply, and the sums to be paid become large, the actual transfer of the precious metals from one individual to another is attended with inconvenience and difficulty, and it is found more convenient to substitute written promises to pay on demand specified quantities of gold. These promises are called bank-notes; and when the person or body issuing them is known to be able to fulfil the pledge, the note will circulate for a long time before it gets into the hands of any person who may wish to make use of the gold it /

represents. These paper representatives supply the place of a certain quantity of gold; and, being much cheaper, a large portion of the expense of a metallic circulation is saved by their employment.

172. As commercial transactions increase, the transfer of bank-notes is, to a considerable extent, superseded by shorter processes. Banks are established, into which all monies are paid, and out of which all payments are made, through written orders called checks, drawn by those who keep accounts with them. In a large capital, each bank receives, through its numerous customers, checks payable by every other; and if clerks were sent round to receive the amount in bank-notes due from each, it would occupy much time, and be attended with some risk and inconvenience.

173. *Clearing house.* In London this is avoided, by making all checks paid in to bankers pass through what is technically called *The Clearing House*. In a large room in Lombard Street, about thirty clerks from the several London bankers take their stations, in alphabetical order, at desks placed round the room; each having a small open box by his side, and the name of the firm to which he belongs in large characters on the wall above his head. From time to time other clerks from every house enter the room, and, passing along, drop into the box the checks due by that firm to the house from which this distributor is sent. The clerk at the table enters the amount of the several checks in a book previously prepared, under the name of the bank to which they are respectively due.

Four o'clock in the afternoon is the latest hour to / which the boxes are open to receive checks; and at a few minutes before that time, some signs of increased activity begin to appear in this previously quiet and business-like scene. Numerous clerks then arrive, anxious to distribute, up to the latest possible moment, the checks which have been paid into the houses of their employers.

At four o'clock all the boxes are removed, and each clerk adds up the amount of the checks put into his box and payable by his own to other houses. He also receives another book from his own house, containing the amounts of the checks which their distributing clerk has put into the box of every other banker. Having compared these, he writes out the balances due to or from his own house, opposite the name of each of the other banks; and having verified this statement by a comparison with the similar list made by the clerks of those houses, he sends to his own bank the general balance resulting from this sheet, the amount of which, if it is due from that to other houses, is sent back in bank-notes.

At five o'clock the *Inspector* takes his seat; when each clerk, who has upon the result of all the transactions a balance to pay to various other houses, pays it to the inspector, who gives a ticket for the amount. The clerks of those houses to whom money is due, then receive the several sums from the inspector, who takes from them a ticket for the amount. Thus the whole of these payments are made by a double system of balance, a very small amount of bank-notes passing from hand to hand, and scarcely any coin.

174. It is difficult to form a satisfactory estimate / of the sums which daily pass through this operation: they fluctuate from two millions to perhaps fifteen. About two millions and a half may possibly be considered as something like an average, requiring for its adjustment, perhaps, £200,000 in bank notes and £20 in specie. By an agreement between the different bankers, all checks which have the name of any firm written across them must pass through the clearing house: consequently, if any such check should be lost, the firm on which it is drawn would refuse to pay it at the counter; a circumstance which adds greatly to the convenience of commerce.

The advantage of this system is such, that two meetings a day have been recently established – one at twelve, the other at three o'clock; but the payment of balances takes place once only, at five o'clock.

If all the private banks kept accounts with the Bank of England, it would be possible to carry on the whole of these transactions with a still smaller quantity of circulating medium.

175. In reflecting on the facility with which these vast transactions are accomplished – supposing, for the sake of argument, that they form only the fourth part of the daily transactions of the whole community – it is impossible not to be struck with the importance of interfering as little as possible with their natural adjustment. Each payment indicates a transfer of property made for the benefit of both parties; and if it were possible, which it is not, to place, by legal or other means, some impediment in the way which only amounted to one-eighth per cent, such a species of friction would produce a useless expenditure of nearly four millions annually: a / circumstance which is deserving the attention of those who doubt the good policy of the expense incurred by using the precious metals for one portion of the currency of the country.

176. One of the most obvious differences between a metallic and a

paper circulation is, that the coin can never, by any panic or national danger, be reduced below the value of bullion in other civilized countries; whilst a paper currency may, from the action of such causes, totally lose its value. Both metallic and paper money, it is true, may be depreciated, but with very different effects.

1. *Depreciation of coin.* The state may issue coin of the same nominal value, but containing only half the original quantity of gold, mixed with some cheap alloy; but every piece so issued bears about with it internal evidence of the amount of the depreciation: it is not necessary that every successive proprietor should analyse the new coin; but a few having done so, its intrinsic worth becomes publicly known. Of course the coin previously in circulation is now more valuable as bullion, and quickly disappears. All future purchases adjust themselves to the new standard, and prices are quickly doubled; but all past contracts also are vitiated, and all persons to whom money is owing, if compelled to receive payment in the new coin, are robbed of one-half of their debt, which is confiscated for the benefit of the debtor.

2. *Depreciation of paper.* The depreciation of paper money follows a different course. If, by any act of the Government paper is ordained to be a legal tender for debts, and, at the same time, ceases to be / exchangeable for coin, those who have occasion to purchase of foreigners, who are not compelled to take the notes, will make some of their payments in gold; and if the issue of paper, unchecked by the power of demanding the gold it represents, be continued, the whole of the coin will soon disappear. But the public, who are obliged to take the notes, are unable, by any internal evidence, to detect the extent of their depreciation; it varies with the amount in circulation, and may go on till the notes shall be worth little more than the paper on which they are printed. During the whole of this time every creditor is suffering to an extent which he cannot measure; and every bargain is rendered uncertain in its advantage, by the continually changing value of the medium through which it is conducted. This calamitous course has actually been run in several countries: in France, it reached nearly its extreme limit during the existence of assignats. We have ourselves experienced some portion of the misery it creates; but by a return to sounder principles, have happily escaped the destruction and ruin which always attends the completion of that career.

177. Every person in a civilized country requires, according to his

station in life, the use of a certain quantity of money, to make the ordinary purchases of the articles which he consumes. The same individual pieces of coin, it is true, circulate again and again, in the same district: the identical piece of silver, received by the workman on Saturday night, passing through the hands of the butcher, the baker, and the small tradesman, is, perhaps, given by the latter to the manufacturer in exchange / for his check, and is again paid into the hands of the workman at the end of the succeeding week. Any deficiency in this supply of money is attended with considerable inconvenience to all parties. If it be only in the smaller coins, the first effect is a difficulty in procuring small change; then a disposition in the shopkeepers to refuse change unless a purchase to a certin amount be made; and, finally, a premium in money will be given for changing the larger denominations of coin.

Thus money itself varies in price, when measured by other money in larger masses: and this effect takes place whether the circulating medium is metallic or of paper. These effects have constantly occurred, and particularly during the late war; and, in order to relieve it, silver tokens for various sums were issued by the Bank of England.

The inconvenience and loss arising from a deficiency of small money fall with greatest weight on the classes whose means are least; for the wealthier buyers can readily procure credit for their small purchases, until their bill amounts to one of the larger coins.

178. As money, when kept in a drawer, produces nothing, few people, in any situation of life, will keep, either in coin or in notes, more than is immediately necessary for their use; when, therefore, there are no profitable modes of employing money, a superabundance of paper will return to the source from whence it issued, and an excess of coin will be converted into bullion and exported.

179. Since the worth of all property is measured by money, it is obviously conducive to the / general welfare of the community, that fluctuations in its value should be rendered as small and as gradual as possible.

The evils which result from sudden changes in the value of money will perhaps become more sensible, if we trace their effects in particular instances. Assuming, as we are quite at liberty to do, an extreme case, let us suppose three persons, each possessing a hundred pounds: one of these, a widow advanced in years, and who, by the advice of her friends, purchases with that sum an annuity of twenty pounds a year during her life: and let the two others be workmen, who, by industry and economy, have each saved a hundred pounds out of their wages; both these latter persons proposing to procure machines for calendering, and to commence that business. One of these invests his money in a savings' bank; intending to make his own calendering machine, and calculating that he shall expend twenty pounds in materials, and the remaining eighty in supporting himself and in paying the workmen who assist him in constructing it. The other workman, meeting with a machine which he can buy for two hundred pounds, agrees to pay for it a hundred pounds immediately, and the remainder at the end of a twelvemonth. Let us now imagine some alteration to take place in the currency, by which it is depreciated one-half: prices soon adjust themselves to the new circumstances, and the annuity of the widow, though nominally of the same amount, will, in reality, purchase only half the quantity of the necessaries of life which it did before. The workman who had placed his money in the savings' bank, having perhaps purchased ten / pounds' worth of materials, and expended ten pounds in labour applied to them, now finds himself, by this alteration in the currency, possessed nominally of eighty pounds, but in reality of a sum which will purchase only half the labour and materials required to finish his machine; and he can neither complete it, from want of capital, nor dispose of what he has already done in its unfinished state for the price it has cost him. In the meantime, the other workman, who had incurred a debt of a hundred pounds in order to complete the purchase of his calendering machine, finds that the payments he receives for calendering, have, like all other prices, doubled, in consequence of the depreciation of the currency; and he has therefore, in fact, obtained his machine for one hundred and fifty pounds. Thus, without any fault or imprudence, and owing to circumstances over which they have no control, the widow is reduced almost to starve; one workman is obliged to renounce, for several years, his hope of becoming a master; and another, without any superior industry or skill, but in fact, from having made, with reference to his circumstances, rather an imprudent bargain, finds himself unexpectedly relieved from half his debt, and the possessor of a valuable source of profit; whilst the former owner of the machine, if he also has invested the money arising from its sale in the savings' bank, finds his property suddenly reduced one-half.

180. These evils, to a greater or less extent, attend every change in the value of the currency; and the importance of preserving it as far as possible unaltered in value, cannot be too strongly impressed upon all classes of the community. /

CHAPTER XV

ON THE INFLUENCE OF VERIFICATION ON PRICE

181. The money price of an article at any given period is usually stated to depend upon the *proportion between the supply and the demand.* The average price of the same article during a long period, is said to depend, ultimately, *on the power of producing and selling it with the ordinary profits of capital.* But these principles, although true in their general sense, are yet so often modified by the influence of others, that it becomes necessary to examine a little into the disturbing forces.

182. With respect to the first of these propositions, it may be observed, that the cost of any article to the purchaser includes, besides the ratio of the supply to the demand, another element, which, though often of little importance, is, in many cases, of great consequence. *The cost, to the purchaser, is the price he pays for any article, added to the cost of verifying the fact of its having that degree of goodness for which he contracts.* In some cases the goodness of the article is evident on mere inspection: and in those cases there is not much difference of price at different shops. The goodness of loaf sugar, for instance, can be discerned almost at a glance; and the consequence is, that the price is so uniform, and the profit upon it so small, that no grocer is at all anxious to sell it; / whilst, on the other hand, tea, of which it is exceedingly difficult to judge, and which can be adulterated by mixture so as to deceive the skill even of a practised eye, has a great variety of different prices, and is that article which every grocer is most anxious to sell to his customers.

The difficulty and expense of verification are, in some instances, so great, as to justify the deviation from well-established principles. Thus it is a general maxim that *Government can purchase any article at a cheaper rate than that at which they can manufacture it themselves.* But it has nevertheless been considered more economical to build extensive flour-mills (such are those at Deptford), and to grind their own corn, than to verify each sack of purchased flour, and to employ persons in devising methods of detecting the new modes of adulteration which might be continually resorted to.

183. Some years since, a mode of preparing old clover and trefoil seeds by a process called *doctoring*, became so prevalent as to excite the attention of the House of Commons. It appeared in evidence before a committee, that the old seed of the white clover was *doctored* by first wetting it slightly, and then drying it with the fumes of burning sulphur; and that the red clover seed had its colour improved by shaking it in a sack with a small quantity of indigo; but this being detected after a time, the *doctors* then used a preparation of logwood, fined by a little copperas, and sometimes by verdigris; thus at once improving the appearance of the old seed, and diminishing, if not destroying, its vegetative power already enfeebled by age. Supposing no injury had resulted to good seed so prepared, it was proved that from the / improved appearance, the market price would be enhanced by this process from five to twenty-five shillings a hundred weight. But the greatest evil arose from the circumstance of these processes rendering old and worthless seed equal in appearance to the best. One witness had tried some *doctored* seed, and found that not above one grain in a hundred grew, and that those which did vegetate died away afterwards; whilst about eighty or ninety per cent of good seed usually grows. The seed so treated was sold to retail dealers in the country, who of course endeavoured to purchase at the cheapest rate, and from them it got into the hands of the farmers; neither of these classes being capable of distinguishing the fraudulent from the genuine seed. Many cultivators, in consequence, diminished their consumption of the article; and others were obliged to pay a higher price to those who had skill to distinguish the mixed seed, and who had integrity and character to prevent them from dealing in it.

184. In the Irish flax trade, a similar example of the high price paid for verification occurs. It is stated in the report of the committee, 'That the natural excellent quality of Irish flax, as contrasted with foreign or British, has been admitted.' Yet from the evidence before that committee it appears that Irish flax sells, in the market, from 1*d*. to 2*d*. per pound less than other flax of equal or inferior quality. Part of this difference of price arises from negligence in its preparation, but a part also from the expense of ascertaining that each parcel is free from useless matter to add to its weight: this

appears from the evidence of Mr J. Corry, who was, during / twenty-seven years, Secretary to the Irish Linen-Board:

The owners of the flax, who are almost always people in the lower classes of life, believe that they can best advance their own interests by imposing on the buyers. Flax being sold by weight, various expedients are used to increase it; and every expedient is injurious, particularly the damping of it; a very common practice, which makes the flax afterwards heat. The inside of every bundle (and the bundles all vary in bulk) is often full of pebbles, or dirt of various kinds, to increase the weight. In this state it is purchased, and exported to Great Britain. The natural quality of Irish flax is admitted to be not inferior to that produced by any foreign country; and yet the flax of every foreign country, imported into Great Britain, obtains a preference amongst the purchasers, because the foreign flax is brought to the British market in a cleaner and more regular state. The extent and value of the sales of foreign flax in Great Britain can be seen by reference to the public accounts; and I am induced to believe, that Ireland, by an adequate extension of her flax tillage, and having her flax markets brought under good regulations, could, without encroaching in the least degree upon the quantity necessary for her home consumption, supply the whole of the demand of the British market, to the exclusion of the foreigners.

185. The lace trade affords other examples; and, in enquiring into the complaints made to the House of Commons by the framework knitters, the committee observe, that, 'It is singular that the grievance most complained of one hundred and fifty years ago, should, in the present improved state of the trade, be the same grievance which is now most complained of: for it appears, by the evidence given before your committee, that *all / the witnesses attribute the decay of the trade more to the making of fraudulent and bad articles, than to the war, or to any other cause.*' And it is shown by the evidence, that a kind of lace called *single-press* was manufactured, which, although good to the eye, became nearly spoiled in washing by the slipping of the threads; that not one person in a thousand could distinguish the difference between *single-press* and *double-press lace*; and that, even workmen and manufacturers were obliged to employ a magnifying glass for that purpose; and that, in another similar article, called *warp lace*, such aid was essential. It was also stated by one witness, that:

The trade had not yet ceased, excepting in those places where the fraud had been discovered; and from those places no orders are now sent for any sort of Nottingham lace, the credit being totally ruined.

186. In the stocking trade similar frauds have been practised. It appeared in evidence, that stockings were made of uniform width from the knee down to the ankle, and being wetted and stretched on frames at the calf, they retained their shape when dry; but that the purchaser could not discover the fraud until, after the first washing, the stockings hung like bags about his ankles.

187. In the watch trade the practice of deceit, in forging the marks and names of respectable makers, has been carried to a great extent both by natives and foreigners; and the effect upon our export trade has been most injurious, as the following extract from the evidence before a committee of the House of Commons will prove: /

Question. How long have you been in the trade?

Answer. Nearly thirty years.

Question. The trade is at present much depressed?

Answer. Yes, sadly.

Question. What is your opinion of the cause of that distress?

Answer. I think it is owing to a number of watches that have been made so exceedingly bad that they will hardly look at them in the foreign markets; all with a handsome outside show, and the works hardly fit for anything.

Question. Do you mean to say, that all the watches made in this country are of that description?

Answer. No; only a number which are made up by some of the Jews, and other low manufacturers. I recollect something of the sort years ago, of a fall-off of the East India work, owing to there being a number of handsome-looking watches sent out, for instance, with hands on and figures, as if they showed seconds, and had not any work regular to show the seconds: the hand went round, but it was not regular.

Question. They had no perfect movements?

Answer. No, they had not; that was a long time since, and we had not any East India work for a long time afterwards.

In the home market, inferior but showy watches are made at a cheap rate, which are not warranted by the maker to go above half an hour; about the time occupied by the Jew pedlar in deluding his country customer.

188. The practice, in retail linen-drapers' shops, of calling certain articles yard wide when the real width is perhaps, only seven-eighths or three-quarters, arose at first from fraud, which being detected, custom was pleaded in its defence: but the result is, that the vender is constantly obliged to measure the width / of his goods in the customer's presence. In all these instances the object of the seller is to

get a higher price than his goods would really produce if their quality were known; and the purchaser, if not himself a skilful judge (which rarely happens to be the case), must pay some person, in the shape of an additional money price, who has skill to distinguish, and integrity to furnish, articles of the quality agreed on. But as the confidence of persons in their own judgement is usually great, large numbers will always flock to the cheap dealer, who thus, attracting many customers from the honest tradesman, obliges him to charge a higher price for his judgement and character than, without such competition, he could afford to do.

189. There are few things which the public are less able to judge of than the quality of drugs; and when these are compounded into medicines it is scarcely possible, even for medical men, to decide whether pure or adulterated ingredients have been employed. This circumstance, concurring with the present injudicious mode of paying for medical assistance, has produced a curious effect on the price of medicines. Apothecaries, instead of being paid for their services and skill, are remunerated by being allowed to place a high charge upon their medicines, which are confessedly of very small pecuniary value. The effect of such a system is an inducement to prescribe more medicine than is necessary; and in fact, even with the present charges, the apothecary, in ninety-nine cases out of a hundred, cannot be fairly remunerated unless the patient either takes, or pays for, more physic than he really requires. The apparent extravagance of the charge of eighteen pence / for a two-ounce phial* of medicine, is obvious to many who do not reflect on the fact that a great part of the charge is, in reality, payment for the exercise of professional skill. As the same charge is made by the apothecary, whether he attends the patient or merely prepares the prescription of a physician, the chemist and druggist soon offered to furnish the same commodity at a greatly diminished price. But the eighteen pence charged by the apothecary might have been fairly divided into two parts, three pence for medicine and bottle, and fifteen pence for attendance. The chemist, therefore, who never *attends* his customers, if he charges only a shilling for the same medicine, realizes a profit of 200 or 300 per cent upon its value. This enormous profit has called into existence a multitude of competitors; and in this instance the

* Apothecaries frequently purchase these phials at the old bottle-warehouses at ten shillings per gross; so that when their servant has washed them, the cost of the phial is nearly one penny.

impossibility of verifying has, in a great measure, counteracted the beneficial effects of competition. The general adulteration of drugs, even at the extremely high price at which they are retailed as medicine, enables those who are supposed to sell them in an unadulterated state to make large profits, whilst the same evil frequently disappoints the expectation, and defeats the skill, of the most eminent physician.

It is difficult to point out a remedy for this evil without suggesting an almost total change in the system of medical practice. If the apothecary were to charge for his visits, and to reduce his medicines / to one-fourth or one-fifth of their present price, he would still have an interest in procuring the best drugs, for the sake of his own reputation or skill. Or if the medical attendant, who is paid more highly for his time, were to have several pupils, he might himself supply the medicines without a specific charge, and his pupils would derive improvement from compounding them, as well as from examining the purity of the drugs he would purchase. The public would gain several advantages by this arrangement. In the first place, it would be greatly for the interest of the medical practitioner to have the best drugs; it would be in his interest also not to give more physic than needful; and it would enable him, through some of his more advanced pupils, to watch more frequently the changes of any malady.

190. There are many articles of hardware which it is impossible for the purchaser to verify at the time of purchase, or even afterwards, without defacing them. Plated harness and coach furniture may be adduced as examples: these are usually of wrought iron covered with silver, owing their strength to the one and a certain degree of permanent beauty to the other metal. Both qualities are, occasionally, much impaired by substituting cast- for wrought-iron, and by plating with soft solder (tin and lead) instead of with hard solder (silver and brass). The loss of strength is the greatest evil in this case; for cast iron, though made for this purpose more tough than usual by careful annealing, is still much weaker than wrought-iron, and serious accidents often arise from harness giving way. In plating with soft solder, a very thin plate of silver is made to cover the / iron, but it is easily detached, particularly by a low degree of heat. Hard soldering gives a better coat of silver, which is very firmly attached, and is not easily injured unless by a very high degree of heat. The inferior can be made to look nearly as well as the better article, and the purchaser can scarcely discover the difference without cutting into it.

191. The principle that *price, at any moment, is dependent on the relation of the supply to the demand*, is true to the full extent only when the whole supply is in the hands of a very large number of small holders, and the demand is caused by the wants of another set of persons, each of whom requires only a very small quantity. And the reason appears to be, that it is only in such circumstances that a uniform average can be struck between the feelings, the passions, the prejudices, the opinions, and the knowledge, of both parties. If the supply, or present stock in hand, be entirely in the possession of one person, he will naturally endeavour to put such a price upon it as shall produce by its sale the greatest quantity of money; but he will be guided in this estimate of the price at which he will sell, both by the knowledge that increased price will cause a diminished consumption, and by the desire to realize his profit before a new supply shall reach the market from some other quarter. If, however, the same stock is in the hands of several dealers, there will be an immediate competition between them, arising partly from their different views of the duration of the present state of supply, and partly from their own peculiar circumstances with respect to the employment of their capital. /

192. The expense of ascertaining that the price charged is that which is legally due is sometimes considerable. The inconvenience which this verification produces in the case of parcels sent by coaches is very great. The time lost in recovering an overcharge generally amounts to so many times the value of the sum recovered, that it is but rarely resorted to. It seems worthy of consideration whether it would not be a convenience to the public if government were to undertake the general conveyance of parcels somewhat on the same system with that on which the post is now conducted. The certainty of their delivery, and the absence of all attempt at overcharge, would render the prohibition of rival carriers unnecessary. Perhaps an experiment might be made on this subject by enlarging the weight allowed to be sent by the two-penny post, and by conveying works in sheets by the general post.

This latter suggestion would be of great importance to literature, and consequently to the circulation of knowledge. As the post-office regulations stand at present, it constantly happens that persons who have an extensive reputation for science, receive by post, from foreign countries, works, or parts of works, for which they are obliged to pay a most extravagant rate of postage, or else refuse to take in some interesting communication. In France and Germany, printed sheets of paper are forwarded by post at a very moderate expense, and it is fit that the science and literature of England should be equally favoured.

193. It is important, if possible, always to connect the name of the workman with the work he has executed: this secures for him the credit or / the blame he may justly deserve; and diminishes, in some cases, the necessity of verification. The extent to which this is carried in literary works, published in America, is remarkable. In the translation of the *Mécanique Celeste*,[a] by Mr Bowditch, not merely the name of the printer, but also those of the compositors, are mentioned in the work.

194. Again, if the commodity itself is of a perishable nature, such, for example, as a cargo of ice imported into the port of London from Norway a few summers since, then time will supply the place of competition; and, whether the article is in the possession of one or of many persons, it will scarcely reach a monopoly price. The history of *cajeput oil* during the last few months, offers a curious illustration of the effect of opinion upon price. In July of last year, 1831, cajeput oil was sold, exclusive of duty, at 7*d*. per ounce. The disease which had ravaged the East was then supposed to be approaching our shores, and its proximity created alarm. At this period, the oil in question began to be much talked of, as a powerful remedy in that dreadful disorder; and in September it rose to the price of 3*s*. and 4*s*. the ounce. In October there were few or no sales: but in the early part of November, the speculations in this substance reached their height, and between the 1st and the 15th it realized the following prices: 3*s*. 9*d*., 5*s*., 6*s*. 6*d*., 7*s*. 6*d*., 8*s*., 9*s*., 10*s*., 10*s*. 6*d*., 11*s*. After 15 November, the holders of cajeput oil were anxious to sell at much lower rates; and in December a fresh arrival was offered by public sale at 5*s*., and withdrawn, being sold afterwards, as it was understood, by private contract, / at 4*s*. or 4*s*. 6*d*. per oz. Since that time, 1*s*. 6*d*. and 1*s*. have been realized; and a fresh arrival, which is daily expected (March, 1832) will probably reduce it below the price of July. Now it is important to notice, that in November, the time of greatest speculation, the quantity in the market was held by *few* persons, and that it frequently changed hands, each holder being

[a] P.S. de Laplace, *Mécanique celeste . . . translated with a commentary by N. Bowditch* (4 vols, Boston, 1829–39).

desirous to realize his profit. The quantity imported since that time has also been considerable.*

195. The effect of the equalization of price by an increased number of dealers, may be observed in the price of the various securities sold at the Stock Exchange. The number of persons who deal in the 3 per cent stock being large, any one desirous of selling can always dispose of his stock at one-eighth per cent under the market price; but those who wish to dispose of bank stock, or of any other securities of more limited circulation, are obliged to make a sacrifice of eight or ten times this amount upon each hundred pounds value.

196. The frequent speculations in oil, tallow, and other commodities, which must occur to the memory of most of my readers, were always founded on the principle of purchasing up all the stock on hand, and agreeing for the purchase of the expected arrivals; thus proving the opinion of capitalists to be, that a larger average price may be procured by the stock being held by few persons. /

* I have understood that the price of camphor, at the same time, suffered similar changes.

[147/8]

CHAPTER XVI

ON THE INFLUENCE OF DURABILITY ON PRICE

197. Having now considered the circumstances that modify what may be called the momentary amount of price, we must next examine a principle which seems to have an effect on its permanent average. *The durability of any commodity influences its cost in a permanent manner.* We have already stated that what may be called the *momentary price* of any commodity depends upon the proportion existing between the supply and demand, and also upon the cost of verification. The *average price*, during a long period, will depend upon the labour required for producing and bringing it to market, as well as upon the average supply and demand; but it will also be influenced by the *durability of the article manufactured.*

Many things in common use are substantially consumed in using: a phosphorus match, articles of food, and a cigar, are examples of this description. Some things after use become inapplicable to their former purposes, as paper which has been printed upon: but it is yet available for the cheesemonger or the trunk-maker. Some articles, as pens, are quickly worn out by use; and some are still valuable after a long continued wear. There are others, few perhaps in number, which never wear out; the harder / precious stones, when well cut and polished, are of this later class: the fashion of the gold or silver mounting in which they are set may vary with the taste of the age, and such ornaments are constantly exposed for sale as second-hand, but the gems themselves, when removed from their supports, are never so considered. A brilliant which has successively graced the necks of a hundred beauties, or glittered for a century upon patrician brows, is weighed by the diamond merchant in the same scale with another which has just escaped from the wheel of the lapidary, and will be purchased or sold by him at the same price per carat. The great mass of commodities is intermediate in its character between these two extremes, and the periods of respective duration are very various. It is evident that the average price of those things which are consumed in the act of using them, can never be less than that of the labour of bringing them to market. They may for a short time be sold for less, but under such circumstances their production must soon cease altogether. On the other hand, if an article never wears out, its price may continue *permanently below* the cost of the labour expended in producing it; and the only consequence will be, that no further production will take place: its price will continue to be regulated by the relation of the supply to the demand; and should that at any after-time rise, for a considerable period, above the cost of production, it will be again produced.

198. Articles become old from actual decay, or the wearing out of their parts; from improved modes of constructing them; or from changes in their form / and fashion, required by the varying taste of the age. In the two latter cases, their utility is but little diminished; and, being less sought after by those who have hitherto employed them, they are sold at a reduced price to a class of society rather below that of their former possessors. Many articles of furniture, such as well-made tables and chairs, are thus found in the rooms of those who would have been quite unable to have purchased them when new; and we find constantly, even in the houses of the more opulent, large looking-glasses which have passed successively through the hands of several possessors, changing only the fashion of their frames; and in some instances even this alteration is omitted, an additional coat of gilding saving them from the character of being second-hand. Thus a taste for luxuries is propagated downwards in society; and, after a short period, the numbers who have acquired new wants become sufficient to excite the ingenuity of the manufacturer to reduce the cost of supplying them, whilst he is himself benefited by the extended scale of demand.

199. There is a peculiarity in looking-glasses with reference to the principle just mentioned. The most frequent occasion of injury to them arises from accidental violence; and the peculiarity is, that, unlike most other articles, when broken they are still of some value. If a large mirror is accidentally cracked, it is immediately cut into two or more smaller ones, each of which may be perfect. If the degree of violence is so great as to break it into many fragments, these smaller pieces may be cut into squares for dressing-glasses; and if the silvering is / injured, it can either be resilvered or used as plate-glass for glazing windows. The addition from our manufactories to the

stock of plate-glass in the country is annually about two hundred and fifty thousand square feet. It would be very difficult to estimate the quantity annually destroyed or exported, but it is probably small; and the effect of these continual additions is seen in the diminished price and increased consumption of the article. Almost all the better order of shop fronts are now glazed with it. If it were quite indestructible, the price would continually diminish; and unless an increased demand arose from new uses, or from a greater number of customers, a single manufactory, unchecked by competition, would ultimately be compelled to shut up, driven out of the market by the permanance of its own productions.

200. The metals are in some degree permanent, although several of them are employed in such forms that they are ultimately lost.

Copper is a metal of which a great proportion returns to use: a part of that employed in sheathing ships and covering houses is lost from corrosion; but the rest is generally remelted. Some is lost in small brass articles, and some is consumed in the formation of salts, Roman vitriol (sulphate of copper), verdigris (acetate of copper), and verditer.

Gold is wasted in gilding and in embroidering; but a portion of this is recovered by burning the old articles. Some portion is lost by the wear of gold, but, upon the whole, it possesses considerable permanence.

Iron. A proportion of this metal is wasted by / oxidation, in small nails, in fine wire; by the wear of tools, and of the tire of wheels, and by the formation of some dyes: but much, both of cast- and of wrought-iron, returns to use.

Lead is wasted in great quantities. Some portion of that which is used in pipes and in sheets for covering roofs returns to the melting-pot; but large quantities are consumed in the form of small shot, or sometimes in that of musket balls, litharge, and red lead, for white and red paints, for glass-making, for glazing pottery, and for sugar of lead (acetate of lead).

Silver is rather a permanent metal. Some portion is consumed in the wear of coin, in that of silver plate, and a portion in silvering and embroidering.

Tin. The chief waste of this metal arises from tinned iron; some is lost in solder and in solutions for the dyers. /

CHAPTER XVII

OF PRICE AS MEASURED BY MONEY

201. The *money price* at which an article sells furnishes us with comparatively little information respecting its value, if we compare distant intervals of time and different countries; for gold and silver, in which price is usually measured, are themselves subject, like all other commodities, to changes in value; nor is there any standard to which these variations can be referred. The average price of a certain quality of different manufactured articles, or of raw produce, has been suggested as a standard; but a new difficulty then presents itself; for the improved methods of producing such articles render their *money price* extremely variable within very limited periods. The annexed table will afford a striking instance of this kind of change within a period of only twelve years. /

Prices of the following articles at Birmingham, in the undermentioned years

Description		*1818*		*1824*		*1828*		*1830*	
		s.	*d.*	*s.*	*d.*	*s.*	*d.*	*s.*	*d.*
Anvils	cwt	25	0	20	0	16	0	13	0
Awls, polished, Liverpool	gross	2	6	2	0	1	6	1	2
Bed-screws, 6 inches long	gross	18	0	15	0	6	0	5	0
Bits, tinned, for bridles	doz.	5	0	5	0	3	3	2	6
Bolts for doors, 6 inches	doz.	6	0	5	0	2	3	1	6
Braces for carpenters, with 12 bits	set	9	0	4	0	4	2	3	5
Buttons, for coats	gross	4	6	6	3	3	0	2	2
Buttons, small, for waistcoats	gross	2	6	2	0	1	2	0	8
Candlesticks, 6 in., brass	pair	2	11	2	0	1	7	1	2
Curry-combs, six barred	doz.	2	9	2	6	1	5	0	11
Frying-pans	cwt	25	0	21	0	18	0	16	0
Gun-locks, single roller	each	6	0	5	2	1	10	1	6
Hammers, shoe, No. 0	doz.	6	9	3	9	3	0	2	9

Description		*1818*		*1824*		*1828*		*1830*	
		s.	*d.*	*s.*	*d.*	*s.*	*d.*	*s.*	*d.*
Hinges, cast-butts, 1 inch	doz.	0	10	0	7½	0	3¾	0	2¾
Knobs, brass, 2 inches for commodes	doz.	4	0	3	6	1	6	1	2
Latches for doors, bright thumb	doz.	2	3	2	2	1	0	0	9
Locks for doors, iron rim, 6 inches	doz.	38	0	32	0	15	0	13	6
Sad-irons and other castings	cwt	22	6	20	0	14	0	11	6
Shovel and tongs, fire-irons	pair	1	0	1	0	0	9	0	6
Spoons, tinned table	gross	17	6	15	0	10	0	7	0
Stirrups, plated	pair	4	6	3	9	1	6	1	1
Trace-chains	cwt	28	0	25	0	19	6	16	6
Trays, japanned tea, 30 inches	each	4	6	3	0	2	0	1	5
Vices for blacksmiths	cwt	30	0	28	0	22	0	19	6
Wire, brass	lb.	1	10	1	4	1	0	0	9
——, iron, No. 6	bund.	16	0	13	0	9	0	7	0

202. I have taken some pains to assure myself of the accuracy of the above table: at different periods of the years quoted the prices may have varied; but / I believe it may be considered as a fair approximation. In the course of my enquiries I have been favoured with another list, in which many of the same articles occur; but in this last instance the prices quoted are separated by an interval of twenty years. It is extracted from the books of a highly respectable house at Birmingham; and the prices confirm the accuracy of the former table, so far as they relate to the articles which are found in that list.

Prices of 1812 and 1832

Description		*1812*		*1832*		*Reduction per cent in price of 1812*
		s.	*d.*	*s.*	*d.*	
Anvils	cwt	25	0	14	0	44
Awls, Liverpool blades	gross	3	6	1	0	71
Candlesticks, iron, plain		3	10¾	2	3½	41
screwed	6 4½	3	9		41	
Bed screws, 6 inch square head	gross	7	6	4	6	40
flat head	gross	8	6	4	8	45
Curry-combs, 6 barred	dozen	4	0½	1	0	75

Description		*1812* s. d.	*1832* s. d.	*Reduction per cent in price of 1812*
Curry-combs, 8 barred	dozen	5 5½	1 5	74
patent, 6 barred	dozen	7 1½	1 5	80
8 barred	dozen	8 6¾	1 10	79
Fire-irons, iron head, No. 1		1 4½	0 7¾	53
No. 2		1 6	0 8½	53
No. 3		1 8¼	0 9½	53
No. 4		1 10½	0 10½	53
Gun-locks, single roller	each	7 2½	1 11	73
Locks, 1¼ brass, port. pad		16 0	2 6	85
2½ inch 3 keyed till-locks	each	2 2	0 9	65
Shoe tacks	gross	5 0	2 0	60
Spoons, tinned, iron table	gross	22 6	7 0	69
Stirrups, com. tinned, 2 bar	dozen	7 0	2 9	61
Trace-chains, iron	cwt	46 9½	15 0	68 /

*Prices of the principal materials, used in mines in Cornwall, at different periods**

ALL DELIVERED AT THE MINES

Description		*1800* s. d.	*1810* s. d.	*1820* s. d.	*1830* s. d.	*1832* s. d.
Coals	wey	81 7	85 5	53 4	51 0	40 0
Timber (balk)	foot	2 0	4 0	1 5	1 0	0 10
(oak)	foot	3 3½	3 0	3 6	3 3	
Ropes	cwt	66 0	84 0	48 6	40 0	40 0
Iron (common bar)	cwt	20 6	14 6	11 0	7 0	6 6
Common castings	cwt	16 0		15 0	8 0	6 6
Pumps	cwt	16*s.* & 17*s.*	17*s.* & 18*s.*	12*s.* & 15*s.*	6 6	6 10
Gunpowder	100 lbs.	114 2	117 6	68 0	52 6	49 0
Candles		9 3	10 0	8 9	5 11	4 10
Tallow	cwt	72 0	84 0	65 8	52 6	43 0
Leather	lb.	2 4	2 3	2 4	2 2	2 1
Blistered steel	cwt			50 0	44 0	38 0
2s. nails	cwt	32 0	28 6	22 0	17 0	16 6

203. I cannot omit availing myself of this opportunity of calling the

* I am indebted to Mr John Taylor for this interesting table.

attention of the manufacturers, merchants, and factors, in *all* our manufacturing and commercial towns, to the great importance, both for their own interests, and for that of the population to which their capital gives employment, of collecting with care such averages from the actual sales registered in their books. Nor, perhaps, would it be without its use to suggest, that such averages would be still more valuable if collected from as many different quarters as possible; that the quantity of the goods from which they are deduced, together with the greatest deviations from the mean, ought to be given; and that if a small committee were to undertake the task, it would give great additional / weight to the information. Political economists have been reproached with too small a use of facts, and too large an employment of theory. If facts are wanting, let it be remembered that the closet-philosopher is unfortunately too little acquainted with the admirable arrangements of the factory; and that no class of persons can supply so readily, and with so little sacrifice of time, the data on which all the reasonings of political economists are founded, as the merchant and manufacturer; and, unquestionably, to no class are the deductions to which they give rise so important. Nor let it be feared that erroneous deductions may be made from such recorded facts: the errors which arise from the absence of facts are far more numerous and more durable than those which result from unsound reasoning respecting true data.

204. The great diminution in price of the articles here enumerated may have arisen from several causes: 1. *The alteration in the value of the currency.* 2. *The increased value of gold in consequence of the increased demand for coin.* The first of these causes may have had some influence; and the second may have had a very small effect upon the two first quotations of prices, but none at all upon the two latter ones. 3. *The diminished rate of profit produced by capital however employed.* This may be estimated by the average price of three per cents at the periods stated. 4. *The diminished price of the raw materials out of which these articles were manufactured.* The raw material is principally brass and iron, and the reduction upon it may, in some measure, be estimated by the diminished price of iron and brass wire, in the cost of which articles, the labour bears a less / proportion than it does in many of the others. 5. *The smaller quantity of raw material employed, and perhaps, in some instances, an inferior quality of workmanship.* 6. *The improved means by which the same effect was produced by diminished labour.*

205. In order to afford the means of estimating the influence of these several causes, the following table is subjoined:

	1812	1818	1824	1828	1830	1832
Average Price of	£ s. d.	£ s. d.	£ s. d.	£ s. d.	£ s. d.	£ s. d.
Gold, per oz.	4 15 6	4 0	3 17 6½	3 17 7	3 17 9½	3 17 10½
Value of currency, per cent	79 5 3	97 6 10	100	100	100	100
Price of 3 per cent consols	59¾	78¼	93⅝	86	89¾	82½
Wheat per quarter	6 5 0	4 1 0	3 2 1	3 11 10	3 14 6	2 19 3
English pig-iron, at Birmingham	7 10 0	6 7 6	6 10 0	5 10 0	4 10 0	—
English bar-iron, at Birmingham	—	10 10 0	9 10 0	7 15 0	6 0 0	5 0 0
Swedish bar-iron, in London, excluding duty of from £4 to £6 10*s*. per ton	16 10 0	17 10 0	14 0 0	14 10 0	13 15 0	13 2 0

As this table, if unaccompanied by any explanation, might possibly lead to erroneous conclusions, I subjoin the following observations, for which I am indebted to the kindness of Mr Tooke, who may yet, I hope, be induced to continue his valuable work on High and Low Prices,[a] through the important period which has elapsed since its publication.

'The table commences with 1812, and exhibits a great falling off in the price of wheat and iron coincidently with a fall in the price of gold, and leading to the inference of cause and effect. Now, as regards wheat, it so happened that in 1812 it reached its / highest price in consequence of a series of bad harvests, when relief by importation was difficult and enormously expensive. In December, 1813, whilst the price of gold had risen to £5, the price of wheat had fallen to 73*s*., or 50 per cent under what it had been in the spring of 1812; proving clearly that the two articles were under the influence of *opposite* causes.

'Again, in 1812, the freight and insurance on Swedish iron were so much higher than at present as to account for nearly the whole of the difference of price: and in 1818 there had been an extensive speculation which had raised the price of all iron, so that a *part* of the subsequent decline was a mere reaction from a previously unfounded elevation. More recently, in 1825, there was a great speculative rise in the article, which served as a strong stimulus to increased production:

[a] T. Tooke, *Thoughts and details on the high and low prices of the last thirty years* (London, 1823).

this, aided by improved power of machinery, has proceeded to such an extent as fully to account for the fall of price.'

To these reflections I will only add, that the result of my own observation leads me to believe that by far the most influential of these causes has been the invention of cheaper modes of manufacturing. The extent to which this can be carried, while a profit can yet be realized at the reduced price, is truly astonishing, as the following fact, which rests on good authority, will prove. Twenty years since, a brass knob for the locks of doors was *made* at Birmingham; the price, at that time, being 13*s*. 4*d*. per dozen. The same article is now *manufactured*, having the same weight of metal, and an equal, or in fact a slightly / superior finish, at 1*s*. 9¼*d*. per dozen. One circumstance which has produced this economy in the *manufacture* is, that the lathe on which these knobs are finished is now turned by a steam-engine; so that the workman, relieved from that labour, can make them twenty times as fast as he did formerly.

206. The difference of price of the same article, when of various dimensions – at different periods in the same country – and in different countries – is curiously contrasted in the annexed table.

Comparative price of plate glass, at the manufactories of London, Paris, Berlin, and Petersburg

DIMENSIONS		LONDON			PARIS		BERLIN	PETERSBURG
Height in inches	*Breadth in inches*	*1771* £ *s.* *d.*	*1794* £ *s.* *d.*	*1832* £ *s.* *d.*	*1825* £ *s.* *d.*	*1835* £ *s.* *d.*	*1828* £ *s.* *d.*	*1825* £ *s.* *d.*
16	16	0 10 3	0 10 1	0 17 6	0 8 7	0 7 6	0 8 11	0 4 10
30	20	1 14 6	2 3 2	2 6 10	1 16 10	1 7 10	1 10 6	1 2 10
50	30	24 2 4	11 5 0	6 12 10	9 0 5	5 0 3	8 13 0	5 15 0
60	40	67 14 10	27 0 0	13 9 6	22 7 5	10 4 3	21 18 0	12 9 0
76	40	—	43 6 0	19 2 9	36 4 5	14 17 5	35 2 11	17 5 0
90	50	—	84 8 0	34 12 9	71 3 8	28 13 4	—	33 18 7
100	75	—	275 0 0	74 5 10	210 13 3	70 9 7	—	—
120	75	—	—	97 15 9	354 3 2	98 3 10	—	—

The price of silvering these plates is twenty per cent on the cost price for English glass; ten per cent on the cost price for Paris plates; and twelve and a half on those of Berlin.

The following table shows the dimensions and price, when silvered, of the largest plates of glass / ever made by the British Plate Glass Company, which are now at their warehouse in London:

Height	*Breadth*	*Price when silvered*
Inches	*Inches*	£ *s. d.*
132	84	200 8 0
146	81	220 7 0
149	84	239 1 6
151	83	239 10 7
160	80	246 15 4

The prices of the largest glass in the Paris lists when silvered, and reduced to English measure, were:

Year	*Inches*	*Inches*	*Price when silvered*
			£ *s. d.*
1825	128	80	629 12 0
1835	128	80	136 19 0

207. If we wish to compare the value of any article at different periods of time, it is clear that neither any one substance, nor even the combination of all manufactured goods, can furnish us with an invariable unit by which to form our scale of estimation. Mr Malthus has proposed for this purpose to consider a day's labour of an agricultural labourer, as the unit to which all value should be referred. Thus, if we wish to compare the value of twenty yards of broad cloth in Saxony at the present time, with that of the same kind and quantity of cloth fabricated in England two centuries ago, we must find the number of days' labour the cloth would have purchased in England at the time mentioned, and compare it with the number of days' labour which the same quantity of cloth will now purchase in Saxony. Agricultural labour appears to have been selected, because it exists in all countries, and employs a large / number of persons, and also because it requires a very small degree of previous instruction. It seems, in fact, to be merely the exertion of a man's physical force; and its value above that of a machine of equal power arises from its portability, and from the facility of directing its efforts to arbitrary and continually fluctuating purposes. It may perhaps be worthy of enquiry, whether a more constant average might not be deduced from combining with this species of labour those trades which require but a moderate exertion of skill, and which likewise exist in all civilized countries, such as those of the blacksmith and carpenter, etc.* In all

* Much information for such an enquiry is to be found, for the particular period to which it refers, in the *Report of the Committee of the House of Commons on Manufacturers' Employment, 2 July, 1830.*

such comparisons there is, however, another element, which, though not essentially necessary, will yet add much to our means of judging. It is an estimate of the quantity of that food on which the labourer usually subsists, which is necessary for his daily support, compared with the quantity which his daily wages will purchase.

208. The existence of a class of middlemen, between small producers and merchants, is frequently advantageous to both parties; and there are certain periods in the history of several manufactures which naturally call that class of traders into existence. There are also times when the advantage ceasing, the custom of employing them also terminates; the middlemen, especially when numerous, as they sometimes are in retail trades, enhancing the price without equivalent good. Thus, in the recent examination / by the House of Commons into the state of the coal trade, it appears that five-sixths of the London public is supplied by a class of middlemen who are called in the trade *Brass plate coal merchants*: these consist principally of merchants' clerks, gentlemen's servants, and others, who have no wharfs of their own, but merely give their orders to some true coal-merchant, who sends in the coals from his wharf: the brass plate coal merchants, of course, receiving a commission for his agency.

209. In Italy this system is carried to a great extent amongst the voituriers, or persons who undertake to convey travellers. There are some possessed of greater fluency and a more persuasive manner, who frequent the inns where the English resort, and who, as soon as they have made a bargain for the conveyance of a traveller, go out amongst their countrymen and procure some other voiturier to do the job for a considerably smaller sum, themselves pocketing the difference. A short time before the day of starting, the contractor appears before his customer in great distress, regretting his inability to perform the journey on account of the dangerous illness of a mother or some relative, and requesting to have his cousin or brother substituted for him. The English traveller rarely fails to acquiesce in this change, and often praises the filial piety of the rogue who has deceived him. /

CHAPTER XVIII

OF RAW MATERIALS

210. Although the cost of any article may be reduced in its ultimate analysis to the quantity of *labour* by which it was produced; yet it is usual, in a certain state of the manufacture of most substances, to call them by the term *raw material*. Thus iron, when reduced from the ore and rendered malleable, is in a state fitted for application to a multitude of useful purposes, and is the *raw material* out of which most of our tools are made. In this stage of its manufacture, but a moderate quantity of *labour* has been expended on the substance; and it becomes an interesting subject to trace the various proportions in which *raw material*, in this sense of the term, and *labour* unite to constitute the value of many of the productions of the arts.

211. Gold leaf consists of a portion of the metal beaten out to so great a degree of thinness, as to allow a greenish-blue light to be transmitted through its pores. About 400 square inches of this are sold, in the form of a small book containing 25 leaves of gold, for 1*s*. 6*d*. In this case, the raw material, or gold, is worth rather less than two-thirds of the manufactured article. In the case of silver leaf, the labour considerably exceeds the value of the material. A book of fifty leaves, which would cover above 1000 square inches, is sold for 1*s*. 3*d*. /

212. We may trace the relative influence of the two causes above referred to, in the prices of fine gold chains made at Venice. The sizes of these chains are known by numbers, the smallest having been (in 1828) No. 1, and the numbers 2, 3, 4, etc., progressively increasing in size. The following table shows the numbers and the prices of those made at that time.* The first column gives the number by which the chain is known; the second expresses the weight in grains of one inch in length of each chain; the third column the number of links in the

* A still finer chain is now manufactured (1832).

same length; and the last expresses the price, in francs worth tenpence each, of a Venetian braccio, or about two English feet of each chain.

Venetian gold chains

No.	*Weight of one inch, in grains*	*Number of links in one inch*	*Price of a Venetian Braccio, equal to two feet 1/8 inch English*
0	·44	98 to 100	60 francs
1	·56	92	40
1½	·77	88	26
2	·99	84	20
3	1·46	72	20
4	1·61	64	21
5	2·09	64	23
6	2·61	60	24
7	3·36	56	27
8	3·65	56	29
9	3·72	56	32
10	5·35	50	34
24	9·71	32	60 /

Amongst these chains, that numbered 0 and that numbered 24 are exactly of the same price, although the quantity of gold in the latter is twenty-two times as much as in the former. The difficulty of making the smallest chain is so great, that the women who make it cannot work above two hours at a time. As we advance from the smaller chain, the proportionate value of the work to the worth of the material becomes less and less, until at the numbers 2 and 3, these two elements of cost balance each other: after which, the difficulty of the work decreases, and the value of the material increases.

213. The quantity of labour expended on these chains is, however, incomparably less than that which is applied in some of the manufactures of iron. In the case of the smallest Venetian chain the value of the labour is not above thirty times that of the gold. The pendulum spring of a watch, which governs the vibrations of the balance, costs at the retail price two pence, and weighs fifteen one-hundredths of a grain, whilst the retail price of a pound of the best

iron, the raw material out of which fifty thousand such springs are made, is exactly the same sum of two pence.[a]

214. The comparative price of labour and of raw material entering into the manufactures of France, has been ascertained with so much care, in a memoir of M. A. M. Héron de Villefosse, *Recherches statistiques, sur les Metaux de France*,* that we shall give an abstract

* Memoires de l'Institut, 1826.

[a] The following appears in the section of additions, pp. [397–9], in the original: The much admired cast-iron ornaments made at Berlin afford another example of the increased value of the raw material, arising from the labour expended upon it. At the establishment of M. Devaranne, a distinguished manufacturer of cast-iron at Berlin, 10,000 of some of the smaller pieces with which the larger ornaments are composed, are required to weigh one pound.

Table of the prices and weights of various ornaments of Berlin iron

Articles	*No. to cwt*	*Price of each* *s.*	*d.*
1. Diadems, 7½ high by 5¼ broad	1100	17	4
2. Bracelets, 7 in. long by 2 in. broad, 72 pieces	2090	9	1
3. Collars, 18 in. long, 1*s*. 6*d*. from 37 to 40 single pieces	2310	6	5
4. Girdle buckles, 3½ in. high, 2¼ broad	2640	2	8
5. Sevigné needles, 2½ in. high by 1½ broad, 11 pieces	9020	4	7
6. Sevigné earrings, 3 in. long, ⅞ broad, 24 pieces, pairs	10,450	5	5
7. Shirt buttons	88,440	0	6

The grey cast-iron used in this manufacture costs about 6*s*. 5*d*. per cwt; therefore the labour bestowed on the material increases its value about 1100 times in the coarser, and nearly 10,000 times in the finer articles. The prices in this table are the retail prices in 1833. In 1827 they were about double, and in 1821 about triple those given above.

The causes of the perfection in this manufacture appears to depend partly on the skill of the moulders, partly on the excellence of the sand, but chiefly it seems to arise from the quality of the iron, which is melted at as low a temperature as possible. Now if this iron when fluid has, like water, a point of maximum density little above its point of solidifying, then if it is poured into a mould at this temperature, it will expand in cooling, and fill up the most minute / traces in the sand; but if it is put in high above that point, the part in contact with the sand will have become solid before the temperatue of the central parts has sunk to that at which it begins to expand, and thus the expansion will probably all take place in the direction of the jet by which the metal entered.

It is much to be regretted that this beautiful manufacture seems to be declining; a circumstance which is attributed mainly to the insufficiency of the protection afforded to new designs. These are originally produced at great expense to the manufacturers, and when once sold to the public are immediately copied by other manufacturers, who, not having contributed to that expense, can easily undersell the original introducers of them. The copies are frequently inferior to the original, and the successive copies becoming gradually deteriorated, have introduced a bad article amongst the smaller manufactures, which greatly injures the foreign trade.

of his results reduced to English measures. The facts respecting the metals relate to the year 1825. /

In France the quantity of raw material which can be purchased for £1, when manufactured into

Silk goods	is worth	£2·37
Broad cloth and woollens		2·15
Hemp and cables		3·94
Linen comprising thread laces		5·00
Cotton goods		2·44

The price of *pig-lead* was £1 1*s*. per cwt; and lead of the value of £1 sterling, became worth, when manufactured into

Sheets or pipes of moderate dimensions	£1·25
White lead	2·60
Ordinary printing characters	4·90
The smallest type	28·30

The price of *copper* was £5 2*s*. per cwt. Copper worth £1 became when manufactured into

Copper sheeting	£1·26
Household utensils	4·77
Common brass pins tinned	2·34
Rolled into plates covered with 1/20 silver	3·56
Woven into metallic cloth, each square inch of which contains 10,000 meshes	52·23

The price of *tin* was £4 12*s*. per cwt. Tin worth £1 when manufactured into

Leaves for silvering glass	became	£1·73
Household utensils		1·85

Quicksilver cost £10 16*s*. per cwt. Quicksilver worth £1 when manufactured into

Vermilion of average quality	became	£1·81 /

Metallic *arsenic* cost £1 4*s*. per cwt. Arsenic worth £1 when manufactured into

White oxide of arsenic	became	£1·83
Sulphuret (orpiment)		4·26

The price of *cast-iron* was 8*s*. per cwt. Cast-iron worth £1 when manufactured into

Household utensils	became	£2·00
Machinery		4·00
Ornamental, as buckles, etc		45·00
Bracelets, figures, buttons, etc.		147·00

Bar-iron cost £1 6*s*. per cwt. Bar-iron worth £1 when manufactured into

Agricultural instruments	became	£3·57
Barrels, musket		9·10
Barrels of double-barrel guns, twisted and damasked		238·08
Blades of penknives		657·14
razor, cast steel		53·57
sabre, for cavalry, infantry, and artillery, etc.	from 9·25 to	16·07
of table knives		35·70
Buckles of polished steel, used as jewellery		896·66
Clothiers' pins		8·03
Door-latches and bolts	from 4·85 to	8·50
Files, common		2·55
flat, cast steel		20·44
Horseshoes		2·55
Iron, small slit, for nails		1·10
Metallic cloth, iron wire, No. 80		96·71
Needles of various sizes	from 17·33 to	70·85
Reeds for weaving 3-4ths calico		21·87
Saws (frame) of steel		5·12
for wood		14·28
Scissors, finest kind		446·94
Steel, cast		4·28
cast, in sheets		6·25
cemented		2·41
natural		1·42
Sword handles, polished steel		972·82
Tinned iron	from 2·04 to	2·34
Wire, iron	from 2·14 to	10·71

215. The following is stated by M. de Villefosse to be the price of bar-iron at the forges of various countries, in January, 1825.

	per ton £	s.	d.
France	26	10	0
Belgium and Germany	16	14	0
Sweden and Russia, at Stockholm and St Petersburg	13	13	0
England, at Cardiff	10	1	0
The price of the article in 1832 was	5	0	0

M. De Villefosse states, that in France bar-iron, made as it usually is with charcoal, costs three times the price of the cast-iron out of which it is made; whilst in England, where it is usually made with coke, the cost is only twice the price of cast-iron.

216. The present price (1832) of *lead* in England is £13 per ton, and the worth of £1 of it manufactured into

Milled sheet lead	becomes	£1·08

The present price of cake *copper* is £84 per ton, and the worth of £1 of it manufactured into

Sheet copper	becomes	£1·11 /

CHAPTER XIX

ON THE DIVISION OF LABOUR

217. Perhaps the most important principle on which the economy of a manufacture depends, is the *division of labour* amongst the persons who perform the work. The first application of this principle must have been made in a very early stage of society; for it must soon have been apparent, that a larger number of comforts and conveniences could be acquired by each individual, if one man restricted his occupation to the art of making bows, another to that of building houses, a third boats, and so on. This division of labour into trades was not, however, the result of an opinion that the general riches of the community would be increased by such an arrangement; but it must have arisen from the circumstance of each individual so employed discovering that he himself could thus make a greater profit of his labour than by pursuing more varied occupations. Society must have made considerable advances before this principle could have been carried into the workshop; for it is only in countries which have attained a high degree of civilization, and in articles in which there is a great competition amongst the producers, that the most perfect system of the division of labour is to be observed. The various principles on which the advantages of this system depend, have been much / the subject of discussion amongst writers on political economy; but the relative importance of their influence does not appear, in all cases, to have been estimated with sufficient precision. It is my intention, in the first instance, to state shortly those principles, and then to point out what appears to me to have been omitted by those who have previously treated the subject.

218. 1. *Of the time required for learning.* It will readily be admitted, that the portion of time occupied in the acquisition of any art will depend on the difficulty of its execution; and that the greater the number of distinct processes, the longer will be the time which the apprentice must employ in acquiring it. Five or seven years have been

adopted, in a great many trades, as the time considered requisite for a lad to acquire a sufficient knowledge of his art, and to enable him to repay by his labour, during the latter portion of his time, the expense incurred by his master at its commencement. If, however, instead of learning *all* the different processes for making a needle, for instance, his attention be confined to one operation, the portion of time consumed unprofitably at the commencement of his apprenticeship will be small, and all the rest of it will be beneficial to his master: and, consequently, if there be any competition amongst the masters, the apprentice will be able to make better terms, and diminish the period of his servitude. Again, the facility of acquiring skill in a single process, and the early period of life at which it can be made a source of profit, will induce a greater number of parents to bring up their children to it; and from this circumstance also, the number / of workmen being increased, the wages will soon fall.

219. 2. *Of waste of materials in learning.* A certain quantity of material will, in all cases, be consumed unprofitably, or spoiled by every person who learns an art; and as he applies himself to each new process, he will waste some of the raw material, or of the partly manufactured commodity. But if each man commit this waste in acquiring successively every process, the quantity of waste will be much greater than if each person confine his attention to one process; in this view of the subject, therefore, the division of labour will diminish the price of production.

220. 3. Another advantage resulting from the division of labour is, *the saving of that portion of time which is always lost in changing from one occupation to another.* When the human hand, or the human head, has been for some time occupied in any kind of work, it cannot instantly change its employment with full effect. The muscles of the limbs employed have acquired a flexibility during their exertion, and those not in action a stiffness during rest, which renders every change slow and unequal in the commencement. Long habit also produces in the muscles exercised a capacity for enduring fatigue to a much greater degree than they could support under other circumstances. A similar result seems to take place in any change of mental exertion; the attention bestowed on the new subject not being so perfect at first as it becomes after some exercise.

221. 4. *Change of tools.* The employment of / different tools in the successive processes is another cause of the loss of time in changing from one operation to another. If these tools are simple, and the

change is not frequent, the loss of time is not considerable; but in many processes of the arts the tools are of great delicacy, requiring accurate adjustment every time they are used; and in many cases the time employed in adjusting bears a large proportion to that employed in using the tool. The sliding-rest, the dividing and the drilling-engine, are of this kind; and hence, in manufactories of sufficient extent, it is found to be good economy to keep one machine constantly employed in one kind of work: one lathe, for example, having a screw motion to its sliding-rest along the whole length of its bed, is kept constantly making cylinders; another, having a motion for equalizing the velocity of the work at the point at which it passes the tool, is kept for facing surfaces; whilst a third is constantly employed in cutting wheels.

222. 5. *Skill acquired by frequent repetition of the same processes.* The constant repetition of the same process necessarily produces in the workman a degree of excellence and rapidity in his particular department, which is never possessed by a person who is obliged to execute many different processes. This rapidity is still further increased from the circumstance that most of the operations in factories, where the division of labour is carried to a considerable extent, are paid for as piece-work. It is difficult to estimate in numbers the effect of this cause upon production. In nail-making, Adam Smith has stated, that it is almost three to one; for, he / observes, that a smith accustomed to make nails, but whose whole business has not been that of a nailer, can make only from eight hundred to a thousand per day; whilst a lad who had never exercised any other trade, can make upwards of two thousand three hundred a day.

223. In different trades, the economy of production arising from the last-mentioned cause will necessarily be different. The case of nail-making is, perhaps, rather an extreme one. It must, however, be observed, that, in one sense, this is not a permanent source of advantage; for, though it acts at the commencement of an establishment, yet every month adds to the skill of the workmen; and at the end of three or four years they will not be very far behind those who have never practised any other branch of their art. Upon an occasion when a large issue of bank-notes was required, a clerk at the Bank of England signed his name, consisting of seven letters, including the initial of his Christian name, five thousand three hundred times during eleven working hours, besides arranging the notes he had signed in parcels of fifty each.

224. 6. *The division of labour suggests the contrivance of tools and machinery to execute its processes.* When each processes, by which any article is produced, is the sole occupation of one individual, his whole attention being devoted to a very limited and simple operation, improvements in the form of his tools, or in the mode of using them, are much more likely to occur to his mind, than if it were distracted by a greater variety of circumstances. Such an improvement in the tool is generally the first step / towards a machine. If a piece of metal is to be cut in a lathe, for example, there is one particular angle at which the cutting-tool must be held to insure the cleanest cut; and it is quite natural that the idea of fixing the tool at that angle should present itself to an intelligent workman. The necessity of moving the tool slowly, and in a direction parallel to itself, would suggest the use of a screw, and thus arises the sliding-rest. It was probably the idea of mounting a chisel in a frame, to prevent its cutting too deeply, which gave rise to the common carpenter's plane. In cases where a blow from a hammer is employed, experience teaches the proper force required. The transition from the hammer held in the hand to one mounted upon an axis, and lifted regularly to a certain height by some mechanical contrivance, requires perhaps a greater degree of invention than those just instanced; yet it is not difficult to perceive, that, if the hammer always falls from the same height, its effect must be always the same.

225. When each process has been reduced to the use of some simple tool, the union of all these tools, actuated by one moving power, constitutes a machine. In contriving tools and simplifying processes, the operative workmen are, perhaps, most successful; but it requires far other habits to combine into one machine these scattered arts. A previous education as a workman in the peculiar trade, is undoubtedly a valuable preliminary; but in order to make such combinations with any reasonable expectation of success, an extensive knowledge of machinery, and the power of making mechanical drawings, are essentially / requisite. These accomplishments are now much more common than they were formerly; and their absence was, perhaps, one of the causes of the multitude of failures in the early history of many of our manufactures.

226. Such are the principles usually assigned as the causes of the advantage resulting from the division of labour. As in the view I have taken of the question, the most important and influential cause has been altogether unnoticed, I shall restate those principles in the words of Adam Smith:

The great increase in the quantity of work, which, in consequence of the division of labour, the same number of people are capable of performing, is owing to three different circumstances: first, to the increase of dexterity in every particular workman; secondly, to the saving of time, which is commonly lost in passing from one species of work to another; and, lastly, to the invention of a great number of machines which facilitate and abridge labour, and enable one man to do the work of many.[a]

Now, although all these are important causes, and each has its influence on the result; yet it appears to me, that any explanation of the cheapness of manufactured articles, as consequent upon the division of labour, would be incomplete if the following principle were omitted to be stated.

*That the master manufacturer, by dividing the work to be executed into different processes, each requiring different degrees of skill or of force, can purchase exactly that precise quantity of both which is necessary for each process; whereas, if the whole work were executed by one workman, that person must possess / sufficient skill to perform the most difficult, and sufficient strength to execute the most laborious, of the operations into which the art is divided.**

227. As the clear apprehension of this principle, upon which a great part of the economy arising from the division of labour depends, is of considerable importance, it may be desirable to point out its precise and numerical application in some specific manufacture. The art of making needles is, perhaps, that which I should have selected for this illustration, as comprehending a very large number of processes remarkably different in their nature; but the less difficult art of pin-making, has some claim to attention, from its having been used by Adam Smith; and I am confirmed in the choice of it, by the circumstance of our possessing a very accurate and minute description of that art, as practised in France above half a century ago.

228. *Pin-making.* In the manufacture of pins in England the following processes are employed:

1. *Wire-drawing.* (*a*) The brass wire used for making pins is purchased by the manufacturer in coils of about twenty-two inches in

* I have already stated that this principle presented itself to me after a personal examination of a number of manufactories and workshops devoted to different purposes; but I have since found that it had been distinctly pointed out, in the work of Gioja, *Nuovo Prospetto delle Scienze Economiche*, 6 tom. 4to. Milano, 1815, tom. i. capo iv.

[a] Adam Smith, *The wealth of nations* (1776), Vol. 1, Book 1, Chap. 1.

diameter, each weighing about thirty-six pounds. (*b*) The coils are wound off into smaller ones of about six inches in diameter, and between one and two pounds' weight. (*c*) The diameter of this wire is now reduced, by / drawing it repeatedly through holes in steel plates, until it becomes of the size required for the sort of pins intended to be made. During this process the wire is hardened, and to prevent its breaking, it must be annealed two or three times, according to the diminution of diameter required. (*d*) The coils are then soaked in sulphuric acid, largely diluted with water, in order to clean them, and are then beaten on stone, for the purpose of removing any oxidated coating which may adhere to them. These operations are usually performed by men, who draw and clean from thirty to thirty-six pounds of wire a day. They are paid at the rate of five farthings per pound, and generally earn about 3*s*. 6*d*. per day.

M. Perronnet made some experiments on the extension the wire undergoes in passing through each hole: he took a picce of thick Swedish brass wire, and found

	Feet	*Inches*
Its length to be before drawing	3	8
After passing the first hole	5	5
second hole	7	2
third hole	7	8

It was now annealed, and the length became

After passing the fourth hole	10	8
fifth hole	13	1
sixth hole	16	8
And finally, after passing through six other holes	144	0

The holes through which the wire was drawn were not, in this experiment, of regularly decreasing diameter: it is extremely difficult to make such / holes, and still more to preserve them in their original dimensions.

229. 2. *Straightening the wire.* The coil of wire now passes into the hands of a woman, assisted by a boy or girl. A few nails, or iron pins, not quite in a line, are fixed into one end of a wooden table about twenty feet in length; the end of the wire is passed alternately between these nails, and is then pulled to the other end of the table. The object of this process is to straighten the wire, which had

acquired a considerable curvature in the small coils in which it had been wound. The length thus straightened is cut off, and the remainder of the coil is drawn into similar lengths. About seven nails or pins are employed in straightening the wire, and their adjustment is a matter of some nicety. It seems, that by passing the wire between the first three nails or pins, a bend is produced in an opposite direction to that which the wire had in the coil; this bend, by passing the next two nails, is reduced to another less curved in the first direction, and so on till the curve of the wire may at last be confounded with a straight line.

230. 3. *Pointing.* (*a*) A man next takes about three hundred of these straightened pieces in a parcel, and putting them into a gauge, cuts off from one end, by means of a pair of shears, moved by his foot, a portion equal in length to rather more than six pins. He continues this operation until the entire parcel is reduced into similar pieces. (*b*) The next step is to sharpen the ends: for this purpose the operator sits before a *steel mill*, which is kept rapidly revolving: it consists of a cylinder about six inches in / diameter, and two and a half inches broad, faced with steel, which is cut in the manner of a file. Another cylinder is fixed on the same axis at a few inches distant; the file on the edge of which is of a finer kind, and is used for finishing off the points. The workman now takes up a parcel of the wires between the finger and thumb of each hand, and presses the ends obliquely on the mill, taking care with his fingers and thumbs to make each wire slowly revolve upon its axis. Having thus pointed all the pieces at one end, he reverses them, and performs the same operation on the other. This process requires considerable skill, but it is not unhealthy; whilst the similar process in needle-making is remarkably destructive of health. (*c*) The pieces now pointed at both ends, are next placed in gauges, and the pointed ends are cut off, by means of shears, to the proper length of which the pins are to be made. The remaining portions of the wire are now equal to about four pins in length, and are again pointed at each end, and their lengths again cut off. This process is repeated a third time, and the small portion of wire left in the middle is thrown amongst the waste, to be melted along with the dust arising from the sharpening. It is usual for a man, his wife, and a child, to join in performing these processes; and they are paid at the rate of five farthings per pound. They can point from thirty-four to thirty-six and a half pounds per day, and gain from 6*s*. 6*d*. to 7*s*., which may

be apportioned thus; 5*s*. 6*d*. the man, 1*s*. the woman, 6*d*. to the boy or girl.

231. 4. *Twisting and cutting the heads.* The next process is making the heads. For this purpose / (*a*) a boy takes a piece of wire, of the same diameter as the pin to be headed, which he fixes on an axis that can be made to revolve rapidly by means of a wheel and strap connected with it. This wire is called the mould. He then takes a smaller wire, which having passed through an eye in a small tool held in his left hand, he fixes close to the bottom of the mould. The mould is now made to revolve rapidly by means of the right hand, and the smaller wire coils round it until it has covered the whole length of the mould. The boy now cuts the end of the spiral connected with the foot of the mould, and draws it off. (*b*) When a sufficient quantity of *heading* is thus made, a man takes from thirteen to twenty of these spirals in his left hand, between his thumb and three outer fingers: these he places in such a manner that two turns of the spiral shall be beyond the upper edge of a pair of shears, and with the forefinger of the same hand he feels that only two turns do so project. With his right hand he closes the shears; and the two turns of the spiral being cut off, drop into a basin; the position of the forefinger preventing the heads from flying about when cut off. The workmen who cut the heads are usually paid at the rate of 2½*d*. to 3*d*. per pound for large heads, but a higher price is given for the smaller heading. Out of this they pay the boy who spins the spiral; he receives from 4*d*. to 6*d*. a day. A good workman can cut from six to about thirty pounds of heading per day, according to its size.

232. 5. *Heading.* The process of fixing the head on the body of the pin is usually executed by women and children. Each operator sits before / a small steel stake, having a cavity, into which one half of the intended head will fit; immediately above is a steel die, having a corresponding cavity for the other half of the head: this latter die can be raised by a pedal moved by the foot. The weight of the hammer is from seven to ten pounds, and it falls through a very small space, perhaps from one to two inches. The cavities in the centre of these dies are connected with the edge of a small groove, to admit of the body of the pin, which is thus prevented from being flattened by the blow of the die. (*a*) The operator with his left hand dips the pointed end of the body of a pin into a tray of heads; having passed the point through one of them, he carries it along to the other end with the

forefinger. He now takes the pin in the right hand, and places the head in the cavity of the stake, and, lifting the die with his foot, allows it to fall on the head. This blow tightens the head on the shank, which is then turned round, and the head receives three or four blows on different parts of its circumference. The women and children who fix the heads are paid at the rate of 1*s*. 6*d*. for every twenty thousand. A skilful operator can with great exertion do twenty thousand per day; but from ten to fifteen thousand is the usual quantity: children head a much smaller number; varying, of course, with the degree of their skill. About one per cent of the pins are spoiled in the process; these are picked out afterwards by women, and are reserved, along with the waste from other processes, for the melting-pot. The die in which the heads are struck is varied in form according to the fashion of the time; but the repeated blows to which it is subject render / it necessary that it should be repaired after it has been used for about thirty pounds of pins.

233. 6. *Tinning*. The pins are now fit to be tinned, a process which is usually executed by a man, assisted by his wife, or by a lad. The quantity of pins operated upon at this stage is usually fifty-six pounds. (*a*) They are first placed in a pickle, in order to remove any grease or dirt from their surface, and also to render them rough, which facilitates the adherence of the tin with which they are to be covered. (*b*) They are then placed in a boiler full of a solution of tartar in water, in which they are mixed with a quantity of tin in small grains. In this they are generally kept boiling for about two hours and a half, and are then removed into a tub of water into which some bran has been thrown, for the purpose of washing off the acid liquor. (*c*) They are then taken out, and, being placed in wooden trays, are well shaken in dry bran: this removes any water adhering to them; and by giving the wooden tray a peculiar kind of motion, the pins are thrown up, and the bran gradually flies off, and leaves them behind in the tray. The man who pickles and tins the pins usually gets one penny per pound for the work, and employs himself, during the boiling of one batch of pins, in drying those previously tinned. He can earn about 9*s*. per day; but out of this he pays about 3*s*. for his assistant.

234. 7. *Papering*. The pins come from the tinner in wooden bowls, with the points projecting in all directions: the arranging of them side by side in paper is generally performed by women. (*a*) A woman takes up some, and places them on a comb, / and shaking them, some of the pins fall back into the bowl, and the rest, being caught by their

heads, are detained between the teeth of the comb. (*b*) Having thus arranged them in a parallel direction, she fixes the requisite number between two pieces of iron, having twenty-five small grooves, at equal distances; (*c*) and having previously doubled the paper, she presses it against the points of the pins until they have passed through the two folds which are to retain them. The pins are then relieved from the grasp of the tool, and the process is repeated. A woman gains about 1*s*. 6*d*. per day by papering; but children are sometimes employed, who earn from 6*d*. per day, and upwards.

235. Having thus generally described the various processes of pin-making, and having stated the usual cost of each, it will be convenient to present a tabular view of the time occupied by each process, and its cost, as well as the sums which can be earned by the persons who confine themselves solely to each process. As the rate of wages is itself fluctuating, and as the prices paid and quantities executed have been given only between certain limits, it is not to be expected that this table can represent the cost of each part of the work with the minutest accuracy, nor even that it shall accord perfectly with the prices above given: but it has been drawn up with some care, and will be quite sufficient to serve as the basis of those reasonings which it is meant to illustrate. A table nearly similar will be subjoined, which has been deduced from a statement of M. Perronet, respecting the art of pin-making in France, above seventy years ago. /

English manufacture

236. Pins, *Elevens*, 5546 weigh one pound; *one dozen* = 6932 pins weigh twenty ounces, and require six ounces of paper.

Name of the process	*Workmen*	*Time for making 1 lb of pins*	*Cost of making 1 lb of pins*	*Workman earns per day*		*Price of making each part of a single pin, in millionths of a penny*
		Hours	*Pence*	*s.*	*d.*	
1. Drawing wire (§ 224)	Man	·3636	1·2500	3	3	225
2. Straightening wire (§ 225)	Woman	·3000	·2840	1	0	51
	Girl	·3000	·1420	0	6	26
3. Pointing (§226)	Man	·3000	1·7750	5	3	319

Name of the process	*Workmen*	*Time for making 1 lb of pins*	*Cost of making 1 lb of pins*	*Workman earns per day*	*Price of making each part of a single pin, in millionths of a penny*
		Hours	*Pence*	*s. d.*	
4. Twisting and cutting heads	Boy	·0400	·0147	0 4½	3
(§ 227)	Man	0·0400	·2103	5 4½	38
5. Heading (§ 228)	Woman	4·0000	5·0000	1 3	901
6. Tinning, or whitening	Man	·1071	·6666	6 0	121
(§ 229)	Woman	·1071	·3333	3 0	60
7. Papering (§ 230)	Woman	2·1314	3·1973	1 6	576
		7·6892	12·8732		2320

Number of persons employed: Men, 4; Women, 4; Children, 2. Total, 10.

French manufacture

237. Cost of 12,000 pins, No. 6, each being eight-tenths of an English inch in length – as they were manufactured in France about 1760; with the cost of each operation: deduced from the observations and statement of M. Perronet. /

Name of the process	*Time for making twelve thousand pins*	*Cost of making twelve thousand pins*	*Workman usually earns per day*	*Expense of tools and materials*
	Hours	*Pence*	*Pence*	*Pence*
1. Wire	—	—	—	24·75
2. Straightening and cutting	1·2	·5	4·5	—
3. Coarse pointing	1·2	·625	10·0	—
Turning wheel*	1·2	·875	7·0	—
Fine pointing	·8	·5	9·375	—
Turning wheel	1·2	·5	4·75	—
Cutting off pointed ends	·6	·375	7·5	—
4. Turning spiral	·5	·125	3·0	—
Cutting off heads	·8	·375	5·625	—
Fuel to anneal ditto	—	—	—	·125
5. Heading	12·0	·333	4·25	—

* The great expense of turning the wheel appears to have arisen from the person so occupied being unemployed during half his time, whilst the pointer went to another manufactory.

Name of the process	Time for making twelve thousand pins	Cost of making twelve thousand pins	Workman usually earns per day	Expense of tools and materials
	Hours	Pence	Pence	Pence
6. {Tartar for cleaning	—	—	—	·5
Tartar for whitening	—	—	—	·5
7. Papering	4·8	·5	2·0	—
Paper	—	—	—	1·0
Wear of tools	—	—	—	2·0
	24·3	4·708		

238. It appears from the analysis we have given of the art of pin-making, that it occupies rather more than seven hours and a half of time, for ten different individuals working in succession on the same material, to convert it into a pound of pins; and that the / total expense of their labour, each being paid in the joint ratio of his skill and of the time he is employed, amounts very nearly to 1*s*. 1*d*. But from an examination of the first of these tables, it appears that the wages earned by the persons employed vary from 4½*d*. per day up to 6*s*., and consequently the skill which is required for their respective employments may be measured by those sums. Now it is evident, that if one person were required to make the whole pound of pins, he must have skill enough to earn about 5*s*. 3*d*. per day, whilst he is pointing the wires or cutting off the heads from the spiral coils – and 6*s*. when he is whitening the pins; which three operations together would occupy little more than the seventeenth part of his time. It is also apparent, that during more than one half of his time he must be earning only 1*s*. 3*d*. per day, in putting on the heads; although his skill, if properly employed, would, in the same time, produce nearly five times as much. If, therefore, we were to employ, for all the processes, the man who whitens the pins, and who earns 6*s*. per day, even supposing that he could make the pound of pins in an equally short time, yet we must pay him for his time 46·14 pence, or about 3*s*. 10*d*. *The pins would therefore cost, in making, three times and three quarters as much as they now do by the application of the division of labour.*

The higher the skill required of the workman in any one process of a manufacture, and the smaller the time during which it is employed, so much the greater will be the advantage of separating that process from the rest, and devoting one person's attention entirely to it. Had we selected the art of needle-making / as our illustration, the economy arising from the division of labour would have been still more striking; for the process of tempering the needles requires great skill, attention, and experience, and although from three to four thousand are tempered at once, the workman is paid a very high rate of wages. In another process of the same manufacture, dry-pointing, which also is executed with great rapidity, the wages earned by the workman reach from 7*s*. to 12*s*., 15*s*., and even, in some instances, to 20*s*. per day; whilst other processes are carried on by children paid at the rate of 6*d*. per day.

239. Some further reflections suggested by the preceding analysis, will be reserved until we have placed before the reader a brief description of a machine for making pins, invented by an American. It is highly ingenious in point of contrivance, and, in respect to its economical principles, will furnish a strong and interesting contrast with the manufacture of pins by the human hand. In this machine a coil of brass wire is placed on an axis; one end of this wire is drawn by a pair of rollers through a small hole in a plate of steel, and is held there by a forceps. As soon as the machine is put in action –

1. The forceps draws the wire on to a distance equal in length to one pin: a cutting edge of steel then descends close to the hole through which the wire entered, and severs the piece drawn out.

2. The forceps holding the piece thus separated moves on, till it brings the wire to the centre of the *chuck* of a small lathe, which opens to receive it. Whilst the forceps is returning to fetch another piece of wire, the lathe revolves rapidly, and grinds the / projecting end of the wire upon a steel mill, which advances towards it.

3. After this first or coarse pointing, the lathe stops, and another forceps takes hold of the half-pointed pin, (which is instantly released by the opening of the *chuck*), and conveys it to a similar *chuck* of an adjacent lathe, which receives it, and finishes the pointing on a finer steel mill.

4. This mill again stops, and another forceps removes the pointed pin into a pair of strong steel clams, having a small groove in them by which they hold the pin very firmly. A part of this groove, which terminates at that edge of the steel clams which is intended to form the head of the pin, is made conical. A small round steel punch is now driven forcibly against the end of the wire thus clamped, and the head of the pin is partially formed by compressing the wire into the conical cavity.

5. Another pair of forceps now removes the pin to another pair of clams, and the head of the pin is completed by a blow from a second punch, the end of which is slightly concave. Each pair of forceps returns as soon as it has delivered its burden; and thus there are always five pieces of wire at the same moment in different stages of advance towards a finished pin.

The pins so formed are received in a tray, and whitened and papered in the usual manner. About sixty pins can thus be made by this machine in one minute; but each process occupies exactly the same time.

240. In order to judge of the value of such a machine, compared with hand labour, it would be / necessary to ascertain: 1. The defects to which pins so made are liable. 2. Their advantages, if any, over those made in the usual way. 3. The prime cost of the machine for making them. 4. The expense of keeping it in repair. 5. The expense of moving the machine and of attending to it.

1. Pins made by the machine are more likely to bend, because the head being 'punched up', the wire must be in a soft state to admit of that operation. 2. Pins made by the machine are better than common ones, because they are not subject to losing their heads. 3. With respect to the prime cost of a machine, it would be very much reduced if a large number should be required. 4. With regard to its wear and tear, experience only can decide: but it may be remarked, that the steel clams or dies in which the heads are punched up, will wear quickly unless the wire has been softened by annealing; and that if softened, the bodies of the pins will bend too readily. Such an inconvenience might be remedied, either by making the machine spin the heads and fix them on, or by annealing only that end of the wire which is to become the head of the pin; but this would cause a delay between the operations, since the brass is too brittle, while heated, to bear a blow without crumbling. 5. On comparing the time occupied by the machine with that stated in the analysis, we find that, except in the heading, the human hand is more rapid. Three thousand six hundred pins are pointed by the machine in one hour, whilst a man can point fifteen thousand six hundred in the same time. But in the process of heading, the rapidity of the machine is two and a half times that of the human hand. It / must, however, be observed, that the grinding in the machine does not require the application of a force equal to that of one man; for all the processes are executed at once by the machine, and a single labourer can easily work it. /

[191/2]

CHAPTER XX

ON THE DIVISION OF MENTAL LABOUR

241. We have already mentioned what may, perhaps, appear paradoxical to some of our readers – that the division of labour can be applied with equal success to mental as to mechanical operations, and that it ensures in both the same economy of time. A short account of its practical application, in the most extensive series of calculations ever executed, will offer an interesting illustration of this fact, whilst at the same time it will afford an occasion for showing that the arrangements which ought to regulate the interior economy of a manufactory, are founded on principles of deeper root than may have been supposed, and are capable of being usefully employed in preparing the road to some of the sublimest investigations of the human mind.

242. In the midst of that excitement which accompanied the Revolution of France and the succeeding wars, the ambition of the nation, unexhausted by its fatal passion for military renown, was at the same time directed to some of the nobler and more permanent triumphs which mark the era of a people's greatness – and which receive the applause of posterity long after their conquests have been wrested from them, or even when their existence as a nation may be told only by the page of history. / Amongst their enterprises of science, the French Government was desirous of producing a series of mathematical tables, to facilitate the application of the decimal system which they had so recently adopted. They directed, therefore, their mathematicians to construct such tables, on the most extensive scale. Their most distinguished philosophers, responding fully to the call of their country, invented new methods for this laborious task; and a work, completely answering the large demands of the Government, was produced in a remarkably short period of time. M. Prony, to whom the superintendence of this great undertaking was confided, in speaking of its commencement,

observes:[a] *Je m'y livrai avec toute l'ardeur dont j'étois capable, et je m'occupai d'abord du plan général de l'exécution. Toutes les conditions que j'avois à remplir nécessitoient l'emploi d'un grand nombre de calculateurs; et il me vint bientôt à la pensée d'appliquer à la confection de ces Tables* la division du travail, *dont les Arts de Commerce tirent un parti si avantageux pour réunir à la perfection de main-d'œuvre l'économie de la dépense et du temps.* The circumstance which gave rise to this singular application of the principle of *the division of labour* is so interesting, that no apology is necessary for introducing it from a small pamphlet printed at Paris a few years since, when a proposition was made by the English to the French Government, that the two countries should print these tables at their joint expense.

243. The origin of the idea is related in the following extract:[b] /

C'est à un chapitre d'un ouvrage Anglais,* justement célèbre, (I.) qu'est probablement due l'existence de l'ouvrage dont le gouvernement Britannique veut faire jouir le monde savant:

* *An Enquiry into the Nature and Causes of the Wealth of Nations*, by Adam Smith.

[a] *Translation:* I devoted myself to it with all the enthusiasm I could muster, and I first concerned myself with the general plan of execution. The many conditions which I had to fulfil called for the use of a large number of calculators; and it soon occurred to me that I could use in the making of these tables of *the division of labour*, such as is turned to such good account by the Commercial Arts in the perfection of work and the saving of time.

[b] *Translation:* The work by which the British Government wishes to benefit the world of learning probably owes its existence to a chapter in a justly famous English work:

Here is the anecdote: M. de Prony undertook, on behalf of the Government committees, to produce *logarithmic and trigonometrical tables for the decimal division of the circle, which, not only would leave nothing to be desired concerning accuracy, but would also form the greatest and most imposing monument of calculation ever made, or even thought of.* The logarithms for the numbers 1 to 200,000 were a necessary and imperative supplement to this labour. It was easy for M. de Prony to determine that even with the help of three or four skilled collaborators, the greatest of possible life expectancies would not suffice to fulfil his commitment. He was concerned with this annoying thought when, finding himself at a bookshop, he saw the fine 1776 London edition of Smith in English; he opened the book at random and chanced on the first chapter, which deals with *the division of labour*, and where the manufacture of pins is given as an example. He had hardly perused the first pages before, by some inspiration, he had the idea of *manufacturing* his logarithms like pins. At that time he was studying at the École Polytechnique an area of analysis which was connected with this type of work, *the method of differences*, and its application to *interpolation*. He went off to spend a few days in the country, and returned to Paris with the *manufacturing* plan, which was subsequently executed. He formed two workshops, which worked separately on the same calculations, and made use of reciprocal verification of the results.

Voici l'anecdote: M. de Prony s'était engagé, avec les comités de gouvernement, à composer pour *la division centesimale du cercle, des tables logarithmiques et trigonometriques, qui, non seulement ne laissassent rien à desirer quant à l'exactitude, mais qui formassent le monument de calcul le plus vaste et le plus imposant qui eût jamais été exécuté, ou même conçu.* Les logarithmes des nombres de 1 à 200,000 formaient à ce travail un supplement nécessaire et exigé. Il fut aisé à M. de Prony de s'assurer que même en s'associant trois ou quatre habiles co-opèrateurs, la plus grande durée presumable de sa vie, ne lui suffirai pas pour remplir ses engagements. Il était occupé de cette fâcheuse pensée lorsque, se trouvant devant la boutique d'un marchand de livres, il apperçut la belle edition Anglaise de Smith, donnée a Londres en 1776; il ouvrit le livre au hazard, et tomba sur le premier chapitre, qui traite de *la division du travail*, et où la fabrication des épingles est citée pour exemple. A peine avait-il parcouru les premières pages, que, par une espèce d'inspiration, il conçut l'expédient de mettre ses logarithmes en *manufacture* comme les épingles. Il faisait, en ce moment, à l'école polytechnique, des leçons sur une partie d'analyse liée à ce genre de travail, *la methode des differences*, et ses applications à *l'interpolation*. Il alla passer quelques jours à la campagne, et revint à Paris avec le plan de *fabrication*, qui a été suivi dans l'exécution. Il rassembla deux ateliers, qui faisaient séparément les mêmes calculs, et se servaient de vérification reciproque.*

244. The ancient methods of computing tables were altogether inapplicable to such a proceeding. M. Prony, therefore, wishing to avail himself of all the talent of his country in devising new methods, formed the first section of those who were to take part in this enterprise out of five or six of the most eminent mathematicians in France.

First section. The duty of this first section was to investigate, amongst the various analytical expressions which could be found for the same function, that which was most readily adapted to simple numerical calculation by many individuals employed at the same time. This section had little or nothing to do with the actual numerical work. When its labours were concluded, the formulae on the use of which it had decided, were delivered to the second section.

Second section. This section consisted of seven or eight persons of considerable acquaintance with mathematics: and their duty was to convert into numbers the formulae put into their hands by the first section – an operation of great labour; and then to deliver out these

* *Note sur la publication, proposée par le gouvernement / Anglais des grandes tables logarithmiques et trigonometriques de M. de Prony.* De l'imprimerie de F. Didot, December 1, 1820, p. 7.

formulae to the members of the third section, and receive from them the finished calculations. The members of this second section had certain means of verifying the calculations without the necessity of repeating, or even of examining, the whole of the work done by the third section.

Third section. The members of this section, whose number varied from sixty to eighty, received certain numbers from the second section, and, using nothing more than simple addition and subtraction, / they returned to that section the tables in a finished state. It is remarkable that nine-tenths of this class had no knowledge of arithmetic beyond the two first rules which they were thus called upon to exercise, and that these persons were usually found more correct in their calculations, than those who possessed a more extensive knowlege of the subject.

245. When it is stated that the tables thus computed occupy seventeen large folio volumes, some idea may perhaps be formed of the labour. From that part executed by the third class, which may almost be termed mechanical, requiring the least knowledge and by far the greatest exertions, the first class were entirely exempt. Such labour can always be purchased at an easy rate. The duties of the second class, although requiring considerable skill in arithmetical operations, were yet in some measure relieved by the higher interest naturally felt in those more difficult operations. The exertions of the first class are not likely to require, upon another occasion, so much skill and labour as they did upon the first attempt to introduce such a method; but when the completion of a calculating engine shall have produced a substitute for the whole of the third section of computers, the attention of analysts will naturally be directed to simplifying its application, by a new discussion of the methods of converting analytical formulae into numbers.

246. The proceeding of M. Prony, in this celebrated system of calculation, much resembles that of a skilful person about to construct a cotton or silk mill, or any similar establishment. Having, by his own genius, or through the aid of his friends, / found that some improved machinery may be successfully applied to his pursuit, he makes drawings of his plans of the machinery, and may himself be considered as constituting the first section. He next requires the assistance of operative engineers capable of executing the machinery he has designed, some of whom should understand the nature of the processes to be carried on; and these constitute his second section.

When a sufficient number of machines have been made, a multitude of other persons, possessed of a lower degree of skill, must be employed in using them; these form the third section: but their work, and the just performance of the machines, must be still superintended by the second class.

247. As the possibility of performing arithmetical calculations by machinery may appear to non-mathematical readers to be rather too large a postulate, and as it is connected with the subject of the *division of labour*, I shall here endeavour, in a few lines, to give some slight perception of the manner in which this can be done – and thus to remove a small portion of the veil which covers that apparent mystery.

248. *That nearly all tables of numbers which follow any law, however complicated, may be formed, to a greater or less extent, solely by the proper arrangement of the successive addition and subtraction of numbers befitting each table*, is a general principle which can be demonstrated to those only who are well acquainted with mathematics; but the mind, even of the reader who is but very slightly acquainted with that science, will readily conceive that it is not impossible, by attending to the following example.

The subjoined table is the beginning of one in / very extensive use, which has been printed and reprinted very frequently in many countries, and is called a *table of square numbers*.

Terms of the table	*A Table*	*B First difference*	*C Second difference*
1	1		
		3	
2	4		2
		5	
3	9		2
		7	
4	16		2
		9	
5	25		2
		11	
6	36		2
		13	
7	49		

Any number in the table, column A, may be obtained, by multiplying the number which expresses the distance of that term from the commencement of the table by itself; thus, 25 is the fifth term from the beginning of the table, and 5 multiplied by itself, or by 5, is equal to 25. Let us now subtract each term of this table from the next succeeding term, and place the results in another column (B), which may be called first difference column. If we again subtract each term of this first difference from the succeeding term, we find the result is always the number 2, (column C); and that the same number will always recur in that column, which may be called the second difference, will appear to any person who / takes the trouble to carry on the table a few terms further. Now when once this is admitted, it is quite clear that, provided the first term (1) of the table, the first term (3) of the first differences, and the first term (2) of the second or constant difference, are originally given, we can continue the table of square numbers to any extent, merely by addition: for the series of first differences may be formed by repeatedly adding the constant difference (2) to (3) the first number in column B, and we then have the series of numbers, 3, 5, 6, etc.: and again, by successively adding each of these to the first number (1) of the table, we produce the square numbers.

249. Having thus, I hope, thrown some light upon the theoretical part of the question, I shall endeavour to show that the mechanical execution of such an engine, as would produce this series of numbers, is not so far removed from that of ordinary machinery as might be conceived.* Let the reader imagine three clocks, placed on a table side by side, each having only one hand, and each having a thousand divisions instead of twelve hours marked on the face; and every time a string is pulled, let them strike on a bell the numbers of the divisions to which their hands point. Let him further suppose that two of the clocks, for the sake of distinction called B and C, have some mechanism by which the clock C advances the hand of the clock B

* Since the publication of the second edition of this work, one portion of the engine which I have been constructing for some years past has been put together. It calculates, in three columns, a table with its first and second differences. Each column can be expressed as far as five figures, so that these fifteen figures constitute about one ninth part of the / larger engine. The ease and precision with which it works, leave no room to doubt its success in the more extended form. Besides tables of squares, cubes, and portions of logarithmic tables, it possesses the power of calculating certain series whose differences are not constant; and it has already tabulated parts of series formed from the following equations:

$$\Delta^3 u_x = \text{units figure of } \Delta u_x$$

$$\Delta^3 u_x = \text{nearest whole no. to } \left(\frac{1}{10{,}000}\,\Delta u_x\right)$$

one division, for / each stroke it makes upon its own bell: and let the clock B by a similar contrivance advance the hand of the clock A one division, for each stroke it makes on its own bell. With such an arrangement, having set the hand of the clock A to the division I, that of B to III, and that of C to II, let the reader imagine the repeating parts of the clocks to be set in motion continually in the following order: viz. – pull the string of clock A; pull the string of clock B; pull the string of clock C.

The table on the following page will then express the series of movements and their results. /

If now only those divisions struck or pointed at by the clock A be attended to and written down, it will be found that they produce the series of the squares of the natural numbers. Such a series could, of course, be carried by this mechanism only so far as the numbers which can be expressed by three figures; but this may be sufficient to give some idea of the construction – and was, in fact, the point to which the first model of the calculating engine, now in progress, extended.

250. We have seen, then, that the effect of the *division of labour*, both in mechanical and in mental operations, is, that it enables us to purchase and apply to each process precisely that quantity of skill and knowledge which is required for it: we avoid employing any part of the time of a man who can get eight or ten shillings a day by his skill in tempering needles, in turning a wheel, which can be done for sixpence a day; and we equally avoid the loss arising from the employment of an accomplished mathematician in performing the lowest processes of arithmetic.

The subjoined is one amongst the series which it has calculated:

0	3,486	42,972
0	4,991	50,532
1	6,907	58,813
14	9,295	67,826
70	12,236	77,602
230	15,741	88,202
495	19,861	99,627
916	24,597	111,928
1,504	30,010	125,116
2,340	36,131	139,272

The general term of this is,

$$u_x = \frac{x \cdot x-1 \cdot x-2}{1 \cdot 2 \cdot 3} + \text{the whole number in } \frac{x}{10} + 10\,\Sigma^3 \left(\text{units figure of } \frac{x \cdot x+1}{2}\right)$$

Repetitions of Process.	Move-ments.	Clock A. Hand set to I.	Clock B. Hand set to III.	Clock C. Hand set to II.
		TABLE.	*First difference.*	*Second difference.*
1	Pull A.	A. strikes 1	. . . .	. . . .
	— B.	The hand is advanced (by B.) 3 divisions . .	B. strikes 3	. . . .
	— C.	. . . .	The hand is advanced (by C.) 2 divisions . .	C. strikes 2
2	Pull A.	A. strikes 4	. . . .	. . . .
	— B.	The hand is advanced (by B.) 5 divisions . .	B. strikes 5	. . . .
	— C.	. . . .	The hand is advanced (by C.) 2 divisions . .	C. strikes 2
3	Pull A.	A. strikes 9	. . . .	. . . .
	— B.	The hand is advanced (by B.) 7 divisions . .	B. strikes 7	. . . .
	— C.	. . . .	The hand is advanced (by C.) 2 divisions . .	C. strikes 2
4	Pull A.	A. strikes 16	. . . .	. . . .
	— B.	The hand is advanced (by B.) 9 divisions . .	B. strikes 9	. . . .
	— C.	. . . .	The hand is advanced (by C.) 2 divisions . .	C. strikes 2
5	Pull A.	A. strikes 25	. . . .	. . . .
	— B.	The hand is advanced (by B.) 11 divisions .	B. strikes 11	. . . .
	— C.	. . . .	The hand is advanced (by C.) 2 divisions . .	C. strikes 2
6	Pull A.	A. strikes 36	. . . .	. . . .
	— B.	The hand is advanced (by B.) 13 divisions .	B. strikes 13	. . . .
	— C.	. . . .	The hand is advanced (by C.) 2 divisions . .	C. strikes 2

251. The *division of labour* cannot be successfully practised unless there exists a great demand for its produce; and it requires a large capital to be employed in those arts in which it is used. In watchmaking it has been carried, perhaps, to the greatest extent. It was stated in evidence before a committee of the House of Commons, that there are a hundred and two distinct branches of this art, to each of which a boy may be put apprentice: and that he only learns his master's department, and is unable, after his apprenticeship has expired, / without subsequent instruction, to work at any other branch. The watch-finisher, whose business is to put together the scattered parts, is the only one, out of the hundred and two persons, who can work in any other department than his own.

252. In one of the most difficult arts, that of mining, great improvements have resulted from the judicious distribution of the duties; and under the arrangments which have gradually been introduced, the whole system of the mine and its government is now placed under the control of the following officers.

1. A manager, who has the general knowledge of all that is to be done, and who may be assisted by one or more skilful persons.

2. Underground captains direct the proper mining operations, and govern the working miners.

3. The purser and book-keeper manage the accounts.

4. The engineer erects the engines, and superintends the men who work them.

5. A chief pitman has charge of the pumps and the apparatus of the shafts.

6. A surface-captain, with assistants, receives the ores raised, and directs the dressing department, the object of which is to render them marketable.

7. The head carpenter superintends many constructions.

8. The foreman of the smiths regulates the ironwork and tools.

9. A materials man selects, purchases, receives and delivers all articles required.

10. The roper has charge of ropes and cordage of all sorts. /

[203/4]

CHAPTER XXI

ON THE COST OF EACH SEPARATE PROCESS IN A MANUFACTURE

253. The great competition introduced by machinery, and the application of the principle of the subdivision of labour, render it necessary for each producer to be continually on the watch, to discover improved methods by which the cost of the article he manufactures may be reduced; and, with this view, it is of great importance to know the precise expense of every process, as well as of the wear and tear of machinery which is due to it. The same information is desirable for those by whom the manufactured goods are distributed and sold; because it enables them to give reasonable answers or explanations to the objections of enquirers, and also affords them a better chance of suggesting to the manufacturer changes in the fashion of his goods, which may be suitable either to the tastes or to the finances of his customers. To the statesman such knowledge is still more important; for without it he must trust entirely to others, and can form no judgement worthy of confidence, of the effect any tax may produce, or of the injury the manufacturer or the country may suffer by its imposition.

254. One of the first advantages which suggests itself as likely to arise from a correct analysis of the expense of the several processes of any manufacture, is the indication which it would furnish of the course / in which improvement should be directed. If a method could be contrived of diminishing by one fourth the time required for fixing on the heads of pins, the expense of making them would be reduced about thirteen per cent; whilst a reduction of one half the time employed in spinning the coil of wire out of which the heads are cut, would scarcely make any sensible difference in the cost of manufacturing of the whole article. It is therefore obvious, that the attention

would be much more advantageously directed to shortening the former than the latter process.

255. The expense of manufacturing, in a country where machinery is of the rudest kind, and manual labour is very cheap, is curiously exhibited in the price of cotton cloth in the island of Java. The cotton, in the seed, is sold by the picul, which is a weight of about 133 lbs. Not above one fourth or one fifth of this weight, however, is cotton: the natives, by means of rude wooden rollers, can only separate about 1¼ lb. of cotton from the seed by one day's labour. A picul of cleansed cotton, therefore, is worth between four and five times the cost of the impure article; and the prices of the same substance, in its different stages of manufacture, are – for one picul:

	Dollars
Cotton in the seed	2 to 3
Clean cotton	10 to 11
Cotton thread	24
Cotton thread dyed blue	35
Good ordinary cotton cloth	50

Thus it appears that the expense of spinning in Java is 117 per cent on the value of the raw material; the expense of dying thread blue is 45 per cent on / its value; and that of weaving cotton thread into cloth 117 per cent on its value. The expense of spinning cotton into a fine thread is, in England, about 33 per cent.*

256. As an example of the cost of the different processes of a manufacture, perhaps an analytical statement of the expense of the volume now in the reader's hands may not be uninteresting; more especially as it will afford an insight into the nature and extent of the taxes upon literature. It is found economical to print it upon paper of a very large size, so that although thirty-two pages, instead of sixteen, are really contained in each sheet, this work is still called octavo.

* These facts are taken from Crawfurd's *Indian Archipelago*.

				£	s.	d.
To printer, for composing (per sheet of 32 pages) £3 1*s*. 10½ sheets				32	0	6
[This relates to the ordinary size of the type used in the volume.]						
To printer for composing small type, as in extracts and contents, extra per sheet, 3*s*. 10*d*.				2	0	3
To printer, for composing table work, extra per sheet, 5*s*. 6*d*.				2	17	9
Average charge for corrections, per sheet, £3 2*s*. 10*d*.				33	0	0
Press work, 3000 being printed off, per sheet, £3 10*s*.				36	15	0
Paper for 3000, at £1 11*s*. 6*d*. per ream, weighing 28 lbs: the duty on paper at 3*d*. per lb. amounts to 7*s*. per ream, so that the 63 reams which are required for the work will cost:						
Paper	77	3	6			
Excise Duty	22	1	0			
Total expense of paper				99	4	6
Total expense of printing and paper				205	18	0 /
Steel-plate for title-page	0	7	6			
Engraving on ditto, Head of Bacon	2	2	0			
Ditto letters	1	1	0			
Total expense of title-page				3	10	6
Printing title-page, at 6*s*. per 100				9	0	0
Paper for ditto, at 1*s*. 9*d*. per 100				2	12	6
Expenses of advertising				40	0	0
Sundries				5	0	0
Total expense in sheets				266	1	0
Cost of a single copy in sheets; 3052 being printed, including the overplus				0	1	9
Extra boarding				0	0	6
Cost of each copy, boarded*				0	2	3

257. This analysis requires some explanation. The printer usually charges for composition by the sheet, supposing the type to be all of one kind; and as this charge is regulated by the size of the letter, on which the quantity in a sheet depends, little dispute can arise after the price is agreed upon. If there are but few extracts, or other parts of the work, which require to be printed in smaller type; or if there are many notes, or several passages in Greek, or in other languages,

* These charges refer to the edition prepared for the public, and do not relate to the large paper copies in the hands of some of the author's friends.

requiring a different type, these are considered in the original contract, and a small additional price per sheet allowed. If there is a large portion of small type, it is better to have a specific additional charge for it per sheet. If any work with irregular lines and many figures, and what / the printers call rules, occurs, it is called table work, and is charged at an advanced price per sheet. Examples of this are frequent in the present volume.[a] If the page consists entirely of figures, as in mathematical tables, which require very careful correction, the charge for composition is usually doubled. A few years ago I printed a table of logarithms, on a large-sized page, which required great additional labour and care from the readers,* in rendering the proofs correct, and for which, although new punches were not required, several new types were prepared, and for which stereotype plates were cast, costing about £2 per sheet. In this case £11 per sheet were charged, although ordinary composition, with the same sized letter, in demy octavo, could have been executed at thirty-eight shillings per sheet: but as the expense was ascertained before commencing the work, it gave rise to no difficulties.

258. The charge for *corrections* and *alterations* is one which, from the difficulty of measuring them, gives rise to the greatest inconvenience, and is as disagreeable to the publisher (if he be the agent between the author and the printer), and to the master printer or his foreman, as it is to the author himself. If the author study economy, he should make the whole of his corrections in the manuscript, and should copy it out fairly: it will then be printed correctly, and he will have little to pay for corrections. But it is scarcely possible to judge of the effect of any passage correctly, without having it set / up in type; and there are few subjects, upon which an author does not find he can add some details or explanation, when he sees his views in print. If, therefore, he wish to save his own labour in transcribing, and to give the last polish to the language, he must be content to accomplish these objects at an increased expense. If the printer possess a sufficient stock of type, it will contribute still more to the convenience of the author to have his whole work put up in what are technically called *slips*,† and then to make all the corrections, and to have as few revises as he can. The

* *Readers* are persons employed to correct the press at the printing office.

† *Slips* are long pieces of paper on which sufficient matter is printed to form, when divided, from two to four pages of text.

[a] *sic.*

present work was set up in slips, but the corrections have been unusually large, and the revises frequent.

259. The press work, or *printing off*, is charged at a price agreed upon for each two hundred and fifty sheets; and any broken number is still considered as two hundred and fifty. When a large edition is required, the price for two hundred and fifty is reduced; thus, in the present volume, two hundred and fifty copies, if printed alone, would have been charged eleven shillings per sheet, instead of 5*s*. 10*d*., the actual charge. The principle of this mode of charging is good, as it obviates all disputes; but it is to be regretted that the custom of charging the same price for any small number as for two hundred and fifty, is so pertinaciously adhered to, that the workmen will not agree to any other terms when only twenty or thirty copies are required, or even when only three / or four are wanted for the sake of some experiment. Perhaps if all numbers above fifty were charged as two hundred and fifty, and all below as for half two hundred and fifty, both parties would derive an advantage.

260. The effect of the excise duty is to render the paper thin, in order that it may weigh little; but this is counteracted by the desire of the author to make his book look as thick as possible, in order that he may charge the public as much as he decently can; and so on that ground alone the duty is of no importance. There is, however, another effect of this duty, which both the public and the author feel; for they pay, not merely the duty which is charged, but also the profit on that duty, which the paper-maker requires for the use of additional capital; and also the profit to the publisher and bookseller on the increased price of the volume.

261. The estimated charge for advertisements is, in the present case, about the usual allowance for such a volume; and, as it is considered that advertisements in newspapers are the most effectual, where the smallest pays a duty of 3*s*. 6*d*., nearly one half of the charge of advertising is a tax.

262. It appears then, that, to an expenditure of £224 necessary to produce the present volume, £42 are added in the shape of a direct tax. Whether the profits arising from such a mode of manufacturing will justify such a rate of taxation, can only be estimated when the returns from the volume are considered, a subject that will be discussed in a subsequent chapter.* It is at present sufficient / to

* Chapter XXXI.

observe, that the tax on advertisements is an impolitic tax when contrasted with that upon paper, and on other materials employed. The object of all advertisements is, by making known articles for sale, to procure for them a better price, if the sale is to be by auction; or a larger extent of sale if by retail dealers. Now the more any article is known, the more quickly it is discovered whether it contributes to the comfort or advantage of the public; and the more quickly its consumption is assured if it be found valuable. It would appear, then, that every tax on communicating information respecting articles which are the subjects of taxation in another shape, is one which must reduce the amount that would have been raised, had no impediment been placed in the way of making known to the public their qualities and their price. /

[211/2]

CHAPTER XXII

ON THE CAUSES AND CONSEQUENCES OF LARGE FACTORIES

263. On examining the analysis which has been given in chapter XIX of the operations in the art of pin-making, it will be observed, that ten individuals are employed in it, and also that the time occupied in executing the several processes is very different. In order, however, to render more simple the reasoning which follows, it will be convenient to suppose that each of the seven processes there described requires an equal quantity of time. This being supposed, it is at once apparent, that, to conduct an establishment for pin-making most profitably, the number of persons employed must be a multiple of ten. For if a person with small means has only sufficient capital to enable him to employ half that number of persons, they cannot each of them constantly adhere to the execution of the same process; and if a manufacturer employs any number not a multiple of ten, a similar result must ensue with respect to some portion of them. The same reflection constantly presents itself on examining any well-arranged factory. In that of Mr Mordan, the patentee of the ever-pointed pencils, one room is devoted to some of the processes by which steel pens are manufactured. Six fly-presses are here constantly at work; in the first a sheet of thin / steel is brought by the workman under the die which at each blow cuts out a flat piece of the metal, having the form intended for the pen. Two other workmen are employed in placing these flat pieces under two other presses, in which a steel chisel cuts the slit. Three other workmen occupy other presses, in which the pieces so prepared receive their semicylindrical form. The longer time required for adjusting the small pieces in the two latter operations renders them less rapid in execution than the first; so that two workmen are fully occupied in slitting, and three in bending the flat pieces, which one man can punch out of the sheet of steel. If, therefore, it were necessary to enlarge this factory, it is clear that

twelve or eighteen presses would be worked with more economy than any number not a multiple of six.

The same reasoning extends to every manufacture which is conducted upon the principle of the *division of labour*, and we arrive at this general conclusion: *When the number of processes into which it is most advantageous to divide it, and the number of individuals to be employed in it, are ascertained, then all factories which do not employ a direct multiple of this latter number, will produce the article at a greater cost.* This principle ought always to be kept in view in great establishments, although it is quite impossible, even with the best division of the labour, to attend to it rigidly in practice. The proportionate number of the persons who possess the greatest skill, is of course to be first attended to. That exact ratio which is more profitable for a factory employing a hundred workmen, may not be quite the best where there are five hundred; / and the arrangements of both may probably admit of variations, without materially increasing the cost of their produce. But it is quite certain that no individual, nor in the case of pin-making could any five individuals, ever hope to compete with an extensive establishment. Hence arises one cause of the great size of manufacturing establishments, which have increased with the progress of civilization. Other circumstances, however, contribute to the same end, and arise also from the same cause – the division of labour.

264. The material out of which the manufactured article is produced, must, in the several stages of its progress, be conveyed from one operator to the next in succession: this can be done at least expense when they are all working in the same establishment. If the weight of the material is considerable, this reason acts with additional force; but even where it is light, the danger arising from frequent removal may render it desirable to have all the processes carried on in the same building. In the cutting and polishing of glass this is the case; whilst in the art of needle-making several of the processes are carried on in the cottages of the workmen. It is, however, clear that the latter plan, which is attended with some advantages to the family of the workmen, can be adopted only where there exists a sure and quick method of knowing that the work has been well done, and that the whole of the materials given out have been really employed.

265. The inducement to contrive machines for any process of manufacture increases with the demand for the article; and the

introduction of machinery, on / the other hand, tends to increase the quantity produced, and to lead to the establishment of large factories. An illustration of these principles may be found in the history of the manufacture of patent net.

The first machines for weaving this article were very expensive, costing from a thousand to twelve or thirteen hundred pounds. The possessor of one of these, though it greatly increased the quantity he could produce, was nevertheless unable, when working eight hours a day, to compete with the old methods. This arose from the large capital invested in the machinery; but he quickly perceived that with the same expense of fixed capital, and a small addition to his circulating capital, he could work the machine during the whole twenty-four hours. The profits thus realized soon induced other persons to direct their attention to the improvement of those machines; and the price was greatly reduced, at the same time that the rapidity of production of thc patent net was increased. But if machines be kept working through the twenty-four hours, it is necessary that some person shall attend to admit the workmen at the time they relieve each other; and whether the porter or other servant so employed admit one person or twenty, his rest will be equally disturbed. It will also be necessary occasionally to adjust or repair the machine; and this can be done much better by a workman accustomed to machine-making, than by the person who uses it. Now, since the good performance and the duration of machines depend to a very great extent upon correcting every shake or imperfection in their parts as soon as they appear, the prompt attention of a workman resident on the spot will / considerably reduce the expenditure arising from the wear and tear of the machinery. But in the case of single lace frame, or a single loom, this would be too expensive a plan. Here then arises another circumstance which tends to enlarge the extent of a factory. It ought to consist of such a number of machines as shall occupy the whole time of one workman in keeping them in order: if extended beyond that number, the same principle of economy would point out the necessity of doubling or tripling the number of machines, in order to employ the whole time of two or three skilful workmen.

266. Where one portion of the workman's labour consists in the exertion of mere physical force, as in weaving and in many similar arts, it will soon occur to the manufacturer, that if that part were executed by a steam-engine, the same man might, in the case of

weaving, attend to two or more looms at once; and, since we already suppose that one or more operative engineers have been employed, the number of his looms may be so arranged that their time shall be fully occupied in keeping the steam-engine and the looms in order. One of the first results will be, that the looms can be driven by the engine nearly twice as fast as before: and as each man, when relieved from bodily labour, can attend to two looms, one workman can now make almost as much cloth as four. This increase of producing power is, however, greater than that which really took place at first; the velocity of some of the parts of the loom being limited by the strength of the thread, and the quickness with which it commences its motion: but an improvement was soon made, by which the motion commenced / slowly, and gradually acquired greater velocity than it was safe to give it at once; and the speed was thus increased from 100 to about 120 strokes per minute.

267. Pursuing the same principles, the manufactory becomes gradually so enlarged, that the expense of lighting during the night amounts to a considerable sum; and as there are already attached to the establishment persons who are up all night, and can therefore constantly attend to it, and also engineers to make and keep in repair any machinery, the addition of an apparatus for making gas to light the factory leads to a new extension, at the same time that it contributes, by diminishing the expense of lighting, and the risk of accidents from fire, to reduce the cost of manufacturing.

268. Long before a factory has reached this extent, it will have been found necessary to establish an accountant's department, with clerks to pay the workmen, and to see that they arrive at their stated times; and this department must be in communication with the agents who purchase the raw produce, and with those who sell the manufactured article.

269. We have seen that the application of the *division of labour* tends to produce cheaper articles; that it thus increases the demand; and gradually, by the effect of competition, or by the hope of increased gain, that it causes large capitals to be embarked in extensive factories. Let us now examine the influence of this accumulation of capital directed to one object. In the first place, it enables the most important principle on which the advantages of the division of labour depends to be carried almost to its extreme limits: not / merely is the precise amount of skill purchased which is necessary for the execution of each process, but throughout every stage – from that in which the

raw material is procured, to that by which the finished produce is conveyed into the hands of the consumer – the same economy of skill prevails. The quantity of work produced by a given number of people is greatly augmented by such an extended arrangement; and the result is necessarily a great reduction in the cost of the article which is brought to market.

270. Amongst the causes which tend to the cheap production of any article, and which are connected with the employment of additional capital, may be mentioned, the care which is taken to prevent the absolute waste of any part of the raw material. An attention to this circumstance sometimes causes the union of two trades in one factory, which otherwise might have been separated.

An enumeration of the arts to which the horns of cattle are applicable, will furnish a striking example of this kind of economy. The tanner who has purchased the raw hides, separates the horns, and sells them to the makers of combs and lanterns. The horn consists of two parts, an outward horny case, and an inward conical substance, somewhat intermediate between indurated hair and bone. The first process consists in separating these two parts, by means of a blow against a block of wood. The horny exterior is then cut into three portions with a frame-saw.

1. The lowest of these, next the root of the horn, after undergoing several processes, by which it is flattened, is made into combs.

2. The middle of the horn, after being flattened by / heat, and having its transparency improved by oil, is split into thin layers, and forms a substitute for glass, in lanterns of the commonest kind.

3. The tip of the horn is used by the makers of knife handles, and of the tops of whips, and for other similar purposes.

4. The interior, or core of the horn, is boiled down in water. A large quantity of fat rises to the surface; this is put aside, and sold to the makers of yellow soap.

5. The liquid itself is used as a kind of glue, and is purchased by cloth dressers for stiffening.

6. The insoluble substance, which remains behind, is then sent to the mill, and, being ground down, is sold to the farmers for manure.

7. Besides these various purposes to which the different parts of the horn are applied, the clippings, which arise in comb making, are sold to the farmer for manure. In the first year after they are spread over the soil they have comparatively little effect, but during the next four or five their efficiency is considerable. The shavings which form the

refuse of the lantern maker, are of a much thinner texture: some of them are cut into various figures and painted, and used as toys; for being hygrometric, they curl up when placed on the palm of a warm hand. But the greater part of these shavings also are sold for manure, and from their extremely thin and divided form, the full effect is produced upon the first crop.

271. Another event which has arisen, in one trade at least, from the employment of large capital, is, that a class of middlemen, formerly interposed between the maker and the merchant, now / no longer exist. When calico was woven in the cottages of the workmen, there existed a class of persons who travelled about and purchased the pieces so made, in large numbers, for the purpose of selling them to the exporting merchant. But the middlemen were obliged to examine every piece, in order to know that it was perfect, and of full measure. The greater number of the workmen, it is true, might be depended upon, but the fraud of a few would render this examination indispensable: for any single cottager, though detected by one purchaser, might still hope that the fact would not become known to all the rest.

The value of character, though great in all circumstances of life, can never be so fully experienced by persons possessed of small capital, as by those employing much larger sums: whilst these larger sums of money for which the merchant deals, render his character for punctuality more studied and known by others. Thus it happens that high character supplies the place of an additional portion of capital; and the merchant, in dealing with the great manufacturer, is saved from the expense of verification, by knowing that the loss, or even the impeachment, of the manufacturer's character, would be attended with greater injury to himself than any profit upon a single transaction could compensate.

272. The amount of well-grounded confidence, which exists in the character of its merchants and manufacturers, is one of the many advantages that an old manufacturing country always possesses over its rivals. To such an extent is this confidence in character carried in England, that, at one of our / largest towns, sales and purchases on a very extensive scale are made daily in the course of business without any of the parties ever exchanging a written document.

273. A breach of confidence of this kind, which might have been attended with very serious embarrassment, occurred in the recent expedition to the mouth of the Niger.

'We brought with us from England,' Mr Lander states, 'nearly a hundred thousand needles of various sizes, and amongst them was a great quantity of *Whitechapel sharps* warranted *superfine, and not to cut in the eye.* Thus highly recommended, we imagined that these needles must have been excellent indeed; but what was our surprise, some time ago, when a number of them which we had disposed of were returned to us, with a complaint that they were all eyeless, thus redeeming with a vengeance the pledge of the manufacturer, "that they would not cut in the eye". On examination afterwards, we found the same fault with the remainder of the "Whitechapel sharps", so that to save our credit we have been obliged to throw them away.' *

274. The influence of established character in producing confidence operated in a very remarkable manner at the time of the exclusion of British manufactures from the continent during the last war. One of our largest establishments had been in the habit of doing extensive business with a house in the centre of Germany; but, on the closing of the / continental ports against our manufactures, heavy penalties were inflicted on all those who contravened the Berlin and Milan decrees. The English manufacturer continued, nevertheless, to receive orders, with directions how to consign them, and appointments for the time and mode of payment, in letters, the handwriting of which was known to him, but which were never signed, except by the christian name of one of the firm, and even in some instances they were without any signature at all. These orders were executed; and in no instance was there the least irregularity in the payments.

275. Another circumstance may be noticed, which to a small extent is more advantageous to large than to small factories. In the export of several articles of manufacture, a drawback is allowed by government, of a portion of the duty paid on the importation of the raw material. In such circumstances, certain forms must be gone through in order to protect the revenue from fraud; and a clerk, or one of the partners, must attend at the custom-house. The agent of the large establishment occupies nearly the same time in receiving a drawback of several thousands, as the smaller exporter does of a few shillings. But if the quantity exported is inconsiderable, the small manufacturer frequently does not find the drawback will repay him for the loss of time.

276. In many of the large establishments of our manufacturing

* Lander's *Journal of an Expedition to the Mouth of the Niger*, vol. ii, p. 42. <R. and J. Lander, *Journal of an expedition to explore the course of the Niger* . . . (3 vols, London, 1832).>

districts, substances are employed which are the produce of remote countries, and which are, in several instances, almost peculiar to a few situations. The discovery of any new locality, / where such articles exist in abundance, is a matter of great importance to any establishment which consumes them in large quantities; and it has been found, in some instances, that the expense of sending persons to great distances, purposely to discover and to collect such produce, has been amply repaid. Thus it has happened, that the snowy mountains of Sweden and Norway, as well as the warmer hills of Corsica, have been almost stripped of one of their vegetable productions, by agents sent expressly from one of our largest establishments for the dying of calicos. Owing to the same command of capital, and to the scale upon which the operations of large factories are carried on, their returns admit of the expense of sending out agents to examine into the wants and tastes of distant countries, as well as of trying experiments, which, although profitable to them, would be ruinous to smaller establishments possessing more limited resources.

These opinions have been so well expressed in the Report of the Committee of the House of Commons on the Woollen Trade, in 1806, that we shall close this chapter with an extract, in which the advantages of great factories are summed up.

Your committee have the satisfaction of seeing, that the apprehensions entertained of factories are not only vicious in principle, but they are practically erroneous; to such a degree, that even the very opposite principles might be reasonably entertained. Nor would it be difficult to prove, that the factories, to a certain extent at least, and in the present day, seem absolutely necessary to the wellbeing of the domestic system; supplying those very particulars wherein the domestic system must be acknowledged to be / inherently defective: for it is obvious, that the little master manufacturers cannot afford, like the man who possesses considerable capital, to try the experiments which are requisite, and incur the risks, and even losses, which almost always occur, in inventing and perfecting new articles of manufacture, or in carrying to a state of greater perfection articles already established. He cannot learn, by personal inspection, the wants and habits, the arts, manufactures, and improvements of foreign countries; diligence, economy, and prudence, are the requisites of his character, not invention, taste, and enterprise; nor would he be warranted in hazarding the loss of any part of his small capital. He walks in a sure road as long as he treads in the beaten track; but he must not deviate into the paths of speculation. The owner of a factory, on the contrary, being commonly possessed of a large capital, and having all his workmen employed under his own immediate superintendence, may make experiments, hazard speculation, invent shorter or better modes of performing old processes, may introduce new articles, and improve and perfect old ones, thus giving the range to his taste and fancy, and, thereby alone enabling our manufacturers to stand the competition with their commercial rivals in other countries. Meanwhile, as is well worthy of remark (and experience abundantly warrants the assertion), many of these new fabrics and inventions, when their success is once established, become general amongst the whole body of manufacturers; the domestic manufacturers themselves thus benefiting, in the end, from those very factories which had been at first the objects of their jealousy. The history of almost all our other manufactures, in which great improvements have been made of late years in some cases at an immense expense, and after numbers of unsuccessful experiments, strikingly illustrates and enforces the above remarks. It is besides an acknowledged fact, that the owners of factories are often amongst the / most extensive purchasers at the halls, where they buy from the domestic clothier the established articles of manufacture, or are able at once to answer a great and sudden order; whilst, at home, and under their own superintendence, they make their fancy goods, and any articles of a newer, more costly, or more delicate quality, to which they are enabled by the domestic system to apply a much larger proportion of their capital. Thus, the two systems, instead of rivalling, are mutual aids to each other; each supplying the other's defects, and promoting the other's prosperity. /

[225/6]

CHAPTER XXIII

ON THE POSITION OF LARGE FACTORIES

277. It is found in every country, that the situation of large manufacturing establishments is confined to particular districts. In the earlier history of a manufacturing community, before cheap modes of transport have been extensively introduced, it will almost always be found that manufactories are placed near those spots in which nature has produced the raw material: especially in the case of articles of great weight, and in those the value of which depends more upon the material than upon the labour expended on it. Most of the metallic ores being exceedingly heavy, and being mixed up with large quantities of weighty and useless materials, must be smelted at no great distance from the spot which affords them: fuel and power are the requisites for reducing them; and any considerable fall of water in the vicinity will naturally be resorted to for aid in the coarser exertions of physical force; for pounding the ore, for blowing the furnaces, or for hammering and rolling out the iron. There are indeed peculiar circumstances which will modify this. Iron, coal, and limestone, commonly occur in the same tracts; but the union of the fuel in the same locality with the ore does not exist with respect to other metals. The tracts generally the most productive of metallic ores are, geologically speaking, different from those affording coal: thus in / Cornwall there are veins of copper and of tin, but no beds of coal. The copper ore, which requires a very large quantity of fuel for its reduction, is sent by sea to the coalfields of Wales, and is smelted at Swansea; whilst the vessels which convey it, take back coals to work the steam-engines for draining the mines, and to smelt the tin, which requires for that purpose a much smaller quantity of fuel than copper.

278. Rivers passing through districts rich in coal and metals, will form the first highroads for the conveyance of weighty produce to stations in which other conveniences present themselves for the further

application of human skill. Canals will succeed, or lend their aid to these; and the yet unexhausted applications of steam and of gas, hold out a hope of attaining almost the same advantages for countries to which nature seemed for ever to have denied them. Manufactures, commerce, and civilization, always follow the line of new and cheap communications. Twenty years ago, the Mississippi poured the vast volume of its waters in lavish profusion through thousands of miles of countries, which scarcely supported a few wandering and uncivilized tribes of Indians. The power of the stream seemed to set at defiance the efforts of man to ascend its course; and, as if to render the task still more hopeless, large trees, torn from the surrounding forests, were planted like stakes in its bottom, forming in some places barriers, in others the nucleus of banks; and accumulating in the same spot, which but for accident would have been free from both, the difficulties and dangers of shoals and of rocks. Four months of incessant toil could scarcely convey a small bark with its worn-out / crew two thousand miles up this stream. The same voyage is now performed in fifteen days by large vessels impelled by steam, carrying hundreds of passengers enjoying all the comforts and luxuries of civilized life. Instead of the hut of the Indian, and the far more unfrequent log house of the thinly scattered settlers – villages, towns, and cities, have arisen on its banks; and the same engine which stems the force of these powerful waters, will probably tear from their bottom the obstructions which have hitherto impeded and rendered dangerous their navigation.*

279. The accumulation of many large manufacturing establishments

* The amount of obstructions arising from the casual fixing of trees in the bottom of the river, may be estimated from the proportion of steamboats destroyed by running upon them. The subjoined statement is taken from the American Almanack for 1832:

'Between the years 1811 and 1831, three hundred and forty-eight steamboats were built on the Mississippi and its tributary streams. During that period a hundred and fifty were lost or worn out.

'Of this hundred and fifty	worn out	63
	lost by snags	36
	burnt	14
	lost by collision	3
	by accidents not ascertained	34'

Thirty-six, or nearly one fourth, being destroyed by accidental obstructions.

Snag is the name given in America to trees which stand nearly upright in the stream, with their roots fixed at the bottom.

It is usual to divide off at the bow of the steamboats a watertight chamber, in order that when a hole is made in it by running against the snags, the water may not enter the rest of the vessel and sink it instantly.

in the same district has a tendency / to bring together purchasers or their agents from great distances, and thus to cause the institution of a public mart or exchange. This contributes to diffuse information relative to the supply of raw materials, and the state of demand for their produce, with which it is necessary manufacturers should be well acquainted. The very circumstance of collecting periodically, at one place, a large number both of those who supply the market and of those who require its produce, tends strongly to check the accidental fluctuations to which a small market is always subject, as well as to render the average of the prices much more uniform.

280. When capital has been invested in machinery, and in buildings for its accommodation, and when the inhabitants of the neighbourhood have acquired a knowledge of the modes of working at the machines, reasons of considerable weight are required to cause their removal. Such changes of position do however occur; and they have been alluded to by the Committee on the Fluctuation of Manufacturers' Employment, as one of the causes interfering most materially with an uniform rate of wages: it is therefore of particular importance to the workmen to be acquainted with the real causes which have driven manufactures from their ancient seats.

'The migration or change of place of any manufacture has sometimes arisen from improvements of machinery not applicable to the spot where such manufacture was carried on, as appears to have been the case with the woollen manufacture, which has in great measure migrated from Essex, Suffolk, and other southern counties, to the northern districts, where coal for the use of the steam-engine is much cheaper. But this change has, in / some instances, been caused or accelerated by the conduct of the workmen, in refusing a reasonable reduction of wages, or opposing the introduction of some kind of improved machinery or process; so that, during the dispute, another spot has in great measure supplied their place in the market. *Any violence used by the workmen against the property of their masters, and any unreasonable combination on their part, is almost sure thus to be injurious to themselves.*'*

281. These removals become of serious consequence when the factories have been long established, because a population commensurate with their wants invariably grows up around them. The combinations in Nottinghamshire, of persons under the name of

* This passage is *not* printed in italics in the original, but it has been thus marked in the above extract, from its importance, and from the conviction that the most extended discussion will afford additional evidence of its truth.

Luddites, drove a great number of lace frames from that district, and caused establishments to be formed in Devonshire. We ought also to observe, that the effect of driving any establishment into a new district, where similar works have not previously existed, is not merely to place it out of the reach of such combinations; but, after a few years, the example of its success will most probably induce other capitalists in the new district to engage in the same manufacture: and thus, although one establishment only should be driven away, the workmen, through whose combination its removal is effected, will not merely suffer by the loss of that portion of demand for their labour which the factory caused; but the value of that labour will itself be reduced by the competition of a new field of production. /

282. Another circumstance which has its influence on this question, is the nature of the machinery. Heavy machinery, such as stamping-mills, steam-engines, etc., cannot readily be moved, and must always be taken to pieces for that purpose; but when the machinery of a factory consists of a multitude of separate engines, each complete in itself, and all put in motion by one source of power, such as that of steam, then the removal is much less inconvenient. Thus, stocking frames, lace machines, and looms, can be transported to more favourable positions, with but a small separation of their parts.

283. It is of great importance that the more intelligent amongst the class of workmen should examine into the correctness of these views; because, without having their attention directed to them, the whole class may, in some instances, be led by designing persons to pursue a course, which, although plausible in appearance, is in reality at variance with their own best interests. I confess I am not without a hope that this volume may fall into the hands of workmen, perhaps better qualified than myself to reason upon a subject which requires only plain common sense, and whose powers are sharpened by its importance to their personal happiness. In asking their attention to the preceding remarks, and to those which I shall offer respecting combinations, I can claim only one advantage over them; namely, that I never have had, and in all human probability never shall have, the slightest pecuniary interest, to influence even remotely, or by anticipation, the judgements I have formed on the facts which have come before me. /

CHAPTER XXIV

ON OVER-MANUFACTURING

284. One of the natural and almost inevitable consequences of competition is the production of a supply much larger than the demand requires. This result usually arises periodically; and it is equally important, both to the masters and to the workmen, to prevent its occurrence, or to foresee its arrival. In situations where a great number of very small capitalists exist – where each master works himself and is assisted by his own family, or by a few journeymen – and where a variety of different articles is produced, a curious system of compensation has arisen which in some measure diminishes the extent to which fluctuations of wages would otherwise reach. This is accomplished by a species of middlemen or factors, persons possessing some capital, who, whenever the price of any of the articles in which they deal is greatly reduced, purchase it on their own account, in the hopes of selling at a profit when the market is better. These persons, in ordinary times, act as salesmen or agents, and make up assortments of goods at the market price, for the use of the home or foreign dealer. They possess large warehouses in which to make up their orders, or keep in store articles purchased during periods of depression; thus acting as a kind of flywheel in equalizing the market price. /

285. The effect of over-manufacturing upon great establishments is different. When an over supply has reduced prices, one of two events usually occurs: the first is a diminished payment for labour; the other is a diminution of the number of hours during which the labourers work, together with a diminished rate of wages. In the former case production continues to go on at its ordinary rate: in the latter, the production itself being checked, the supply again adjusts itself to the demand as soon as the stock on hand is worked off, and prices then regain their former level. The latter course appears, in the first instance, to be the best both for masters and men; but there seems to

be a difficulty in accomplishing this, except where the trade is in few hands. In fact, it is almost necessary, for its success, that there should be a combination amongst the masters or amongst the men; or, what is always far preferable to either, a mutual agreement for their joint interests. Combination amongst the men is difficult, and is always attended with the evils which arise from the ill-will excited against any persons who, in the perfectly justifiable exercise of their judgement, are disposed not to act with the majority. The combination of the masters, on the other hand, is unavailing, unless the whole body of them agree: for if any one master can procure more labour for his money than the rest, he will be able to undersell them.

286. If we look only at the interests of the consumer, the case is different. When too large a supply has produced a great reduction of price, it opens the consumption of the article to a new class, and increases the consumption of those who / previously employed it: it is therefore against the interest of both these parties that a return to the former price should occur. It is also certain, that by the diminution of profit which the manufacturer suffers from the diminished price, his ingenuity will be additionally stimulated; that he will apply himself to discover other and cheaper sources for the supply of his raw material; that he will endeavour to contrive improved machinery which shall manufacture it at a cheaper rate; or try to introduce new arrangements into his factory, which shall render the economy of it more perfect. In the event of his success, by any of these courses or by their joint effects, a real and substantial good will be produced. A larger portion of the public will receive advantage from the use of the article, and they will procure it at a lower price; and the manufacturer, though his profit on each operation is reduced, will yet, by the more frequent returns on the larger produce of his factory, find his real gain at the end of the year, nearly the same as it was before; whilst the wages of the workman will return to their level, and both the manufacturer and the workman will find the demand less fluctuating, from its being dependent on a larger number of customers.

287. It would be highly interesting, if we could trace, even approximately, through the history of any great manufacture, the effects of gluts in producing improvements in machinery, or in methods of working; and if we could show what addition to the annual quantity of goods previously manufactured, was produced by each alteration. It would probably / be found, that *the increased*

quantity manufactured by the same capital, when worked with the new improvement, would produce nearly the same rate of profit as other modes of investment.

Perhaps the manufacture of iron* would furnish the best illustration of this subject; because, by having the actual price of pig and bar iron at the same place and at the same time, the effect of a change in the value of currency, as well as several other sources of irregularity, would be removed.

288. At the present moment, whilst the manufacturers of iron are complaining of the ruinously low price of their produce, a new mode of smelting iron is coming into use, which, if it realizes the statement of the patentees, promises to reduce greatly the cost of production. The improvement consists in heating the air previously to employing it for blowing the furnace. One of the results is, that coal may be used instead of coke; and this, in its turn, diminishes the quantity of limestone which is required for the fusion of the iron stone.

The following statement by the proprietors of the patent is extracted from Brewster's Journal,[a] 1832, p. 349: /

Comparative view of the quantity of materials required at the Clyde iron works to smelt a ton of foundry pig-iron, and of the quantity of foundry pig-iron smelted from each furnace weekly

	Fuel in tons of 20 cwt each cwt 112 lbs	*Iron-stone*	*Lime-stone*	*Weekly produce in pig-iron*
	Tons		*Cwt*	*Tons*
1. With air not heated and coke	7	3¼	15	45
2. With air heated and coke	4¾	3¼	10	60
3. With air heated and coals not coked	2¼	3¼	7½	65

Notes. – 1. To the coals stated in the second and third lines, must be added 5 cwt of small coals, required to heat the air.

2. The expense of the *apparatus* for applying the heated air will be from £200 to £300 per furnace.

* The average price per ton of pig-iron, bar-iron, and coal, together with the price paid for labour at the works, for a long series of years, would be very valuable, and I shall feel much indebted to anyone who will favour me with it for any, even short, period.

[a] *Edinburgh Philosophical Journal.*

3. No coals are now coked at the Clyde iron works; at all the three furnaces the iron is smelted with coals.

4. The three furnaces are blown by a double-powered steam-engine, with a steam cylinder 40 inches in diameter, and a blowing cylinder 80 inches in diameter, which compresses the air so as to carry 2½ lbs per square inch. There are two tuyeres to each furnace. The muzzles of the blowpipes are 3 inches in diameter.

5. The air heated to upwards of 600° of Fahrenheit. It will melt lead at the distance of three inches from the orifice through which it issues from the pipe.

289. The increased effect produced by thus heating the air is by no means an obvious result; and an analysis of its action will lead to some curious views respecting the future application of machinery for blowing furnaces.

Every cubic foot of atmospheric air, driven into a / furnace, consists of two gases*; about one-fifth being oxygen, and four-fifths azote.

According to the present state of chemical knowledge, the oxygen alone is effective in producing heat; and the operation of blowing a furnace may be thus analysed.

1. The air is forced into the furnace in a condensed state, and, immediately expanding, abstracts heat from the surrounding bodies.

2. Being itself of moderate temperature, it would, even without expansion, still require heat to raise it to the temperature of the hot substances to which it is to be applied.

3. On coming into contact with the ignited substances in the furnace, the oxygen unites with them, parting at the same moment with a large portion of its latent heat, and forming compounds which have less specific heat than their separate constituents. Some of these pass up the chimney in a gaseous state, whilst others remain in the form of melted slags, floating on the surface of the iron, which is fused by the heat thus set at liberty.

4. The effects of the azote are precisely similar to the first and second of those above described; it seems to form no combinations, and contributes nothing, in any stage, to augment the heat.

The plan, therefore, of heating the air before driving it into the furnace saves, obviously, the whole of that heat which the fuel must have supplied in raising it from the temperature of the external air up

* The accurate proportions are, by measure, oxygen 21, azote 79.

to that of 600° Fahrenheit; thus rendering the fire more intense, / and the glassy slags more fusible, and perhaps also more effectually decomposing the iron ore. The same quantity of fuel, applied at once to the furnace, would only prolong the duration of its heat, not augment its intensity.

290. The circumstance of so large a portion of the air* driven into furnaces being not merely useless, but acting really as a cooling, instead of a heating, cause, added to so great a waste of mechanical power in condensing it, amounting, in fact, to four-fifths of the whole, clearly shows the defects of the present method, and the want of some better mode of exciting combustion on a large scale. The following suggestions are thrown out as likely to lead to valuable results, even though they should prove ineffectual for their professed object.

291. The great difficulty appears to be to separate the oxygen, which aids combustion, from the azote which impedes it. If either of those gases becomes liquid at a lower pressure than the other, and if those pressures are within the limits of our present powers of compression, the object might be accomplished.

Let us assume, for example, that oxygen becomes liquid under a pressure of 200 atmospheres, whilst azote requires a pressure of 250. Then if atmospheric air be condensed to the two hundredth part of its bulk, the oxygen will be found in a liquid state at / the bottom of the vessel in which the condensation is effected, and the upper part of the vessel will contain only azote in the state of gas. The oxygen, now liquefied, may be drawn off for the supply of the furnace; but as it ought when used, to have a very moderate degree of condensation, its expansive force may be previously employed in working a small engine. The compressed azote also in the upper part of the vessel, though useless for combustion, may be employed as a source of power, and, by its expansion, work another engine. By these means the mechanical force exerted in the original compression would all be restored, except that small part retained for forcing the pure oxygen into the furnace, and the much larger part lost in the friction of the apparatus.

292. The principal difficulty to be apprehended in these operations is

* A similar reasoning may be applied to lamps. An Argand burner, whether used for consuming oil or gas, admits almost an unlimited quantity of air. It would deserve enquiry, whether a smaller quantity might not produce greater light; and, possibly, a different supply furnish more heat with the same expenditure of fuel.

that of *packing* a working piston, so as to bear the pressure of 200 or 300 atmospheres: but this does not seem insurmountable. It is possible also that the chemical combination of the two gases which constitute common air may be effected by such pressures: if this should be the case, it might offer a new mode of manufacturing nitrous or nitric acids. The result of such experiments might take another direction: if the condensation were performed over liquids, it is possible that they might enter into new chemical combinations. Thus, if air were highly condensed in a vessel containing water, the latter might unite with an additional dose of oxygen,* which might afterwards be easily disengaged for the use of the furnace. /

293. A further cause of the uncertainty of the results of such an experiment arises from the possibility that azote may really contribute to the fusion of the mixed mass in the furnace, though its mode of operating is at present unknown. An examination of the nature of the gases issuing from the chimneys of iron-foundries, might perhaps assist in clearing up this point; and, in fact, if such enquiries were also instituted upon the various products of all furnaces, we might expect the elucidation of many points in the economy of the metallurgic art.

294. It is very possible also, that the action of oxygen in a liquid state might be exceedingly corrosive, and that the containing vessels must be lined with platinum or some other substance of very difficult oxydation; and most probably new and unexpected compounds would be formed at such pressures. In some experiments made by Count Rumford in 1797, on the force of fired gunpowder, he noticed a solid compound, which always appeared in the gunbarrel when the ignited powder had no means of escaping; and, in those cases, the gas which escaped on removing the restraining pressure was usually inconsiderable.

295. If the liquefied gases are used, the form of the iron furnace must probably be changed, and perhaps it may be necessary to direct the flame from the ignited fuel upon the ore to be fused, instead of mixing that ore with the fuel itself: by a proper regulation of the blast, an oxygenating or a deoxygenating flame might be procured; and from the intensity of the flame, combined with its chemical agency, we might expect the most refractory / ore to be smelted, and that ultimately the metals at present almost infusible, such as platinum,

* Deutoxide of hydrogen, the oxygenated water of Thénard.

titanium, and others, might be brought into common use, and thus effect a revolution in the arts.

296. Supposing, on the occurrence of a glut, that new and cheaper modes of producing are not discovered, and that the production continues to exceed the demand, then it is apparent that too much capital is employed in the trade; and after a time, the diminished rate of profit will drive some of the manufacturers to other occupations. What particular individuals will leave it must depend on a variety of circumstances. Superior industry and attention will enable some factories to make a profit rather beyond the rest; superior capital in others will enable them, without these advantages, to support competition longer, even at a loss, with the hope of driving the smaller capitalists out of the market, and then reimbursing themselves by an advanced price. It is, however, better for all parties, that this contest should not last long; and it is important, that no artificial restraint should interfere to prevent it. An instance of such restriction, and of its injurious effect, occurs at the port of Newcastle, where a particular Act of Parliament requires that every ship shall be loaded in its turn. The Committee of the House of Commons, in their Report on the Coal Trade, state that,

'Under the regulations contained in this Act, if more ships enter into the trade than can be profitably employed in it, the loss produced by detention in port, and waiting for a cargo, which must consequently take place, instead of falling, as it naturally would, / upon particular ships, and forcing them from the trade, is now divided evenly amongst them; and the loss thus created is shared by the whole number.' – *Report*, p. 6.

297. It is not pretended, in this short view, to trace out all the effects or remedies of over-manufacturing; the subject is difficult, and, unlike some of the questions already treated, requires a combined view of the relative influence of many concurring causes. /

[242/3]

CHAPTER XXV

ENQUIRIES PREVIOUS TO COMMENCING ANY MANUFACTORY

298. There are many enquiries which ought always to be made previous to the commencement of the manufacture of any new article. These chiefly relate to the expense of tools, machinery, raw materials, and all the outgoings necessary for its production; to the extent of demand which is likely to arise; to the time in which the circulating capital will be replaced; and to the quickness or slowness with which the new article will supersede those already in use.

299. The expense of tools and of new machines will be more difficult to ascertain, in proportion as they differ from those already employed; but the variety in constant use in our various manufactories, is such, that few inventions now occur in which considerable resemblance may not be traced to others already constructed. The cost of the raw material is usually less difficult to determine; but cases occasionally arise in which it becomes important to examine whether the supply, at the given price, can be depended upon: for, in the case of a small consumption, the additional demand arising from a factory may produce a considerable temporary rise, though it may ultimately reduce the price. /

300. The quantity of any new article likely to be consumed is a most important subject for the consideration of the projector of a new manufacture. As these pages are not intended for the instruction of the manufacturer, but rather for the purpose of giving a general view of the subject, an illustration of the way in which such questions are regarded by practical men, will, perhaps, be most instructive. The following extract from the evidence given before a Committee of the House of Commons, in the Report on Artizans and Machinery, shows the extent to which articles apparently the most insignificant, are consumed, and the view which the manufacturer takes of them.

The person examined on this occasion was Mr Ostler, a manufacturer of glass beads and other toys of the same substance, from Birmingham. Several of the articles made by him were placed upon the table, for the inspection of the Committee of the House of Commons, which held its meetings in one of the committee-rooms.

Question. Is there any thing else you have to state upon this subject?

Answer. Gentlemen may consider the articles on the table as extremely insignificant; but perhaps I may surprise them a little, by mentioning the following fact. Eighteen years ago, on my first journey to London, a respectable-looking man, in the city, asked me if I could supply him with dolls' eyes; and I was foolish enough to feel half offended; I thought it derogatory to my new dignity as a manufacturer, to make dolls' eyes. He took me into a room quite as wide, and perhaps twice the length of this, and we had just room to walk between stacks, from the floor to the ceiling, of parts of dolls. He / said, 'These are only the legs and arms; the trunks are below.' But I saw enough to convince me, that he wanted a great many eyes; and, as the article appeared quite in my own line of business, I said I would take an order by way of experiment; and he showed me several specimens. I copied the order. He ordered various quantities, and of various sizes and qualities. On returning to the Tavistock Hotel, I found that the order amounted to upwards of £500. I went into the country, and endeavoured to make them. I had some of the most ingenious glass toy-makers in the kingdom in my service; but when I showed it to them, they shook their heads, and said they had often seen the article before, but could not make it. I engaged them by presents to use their best exertions; but after trying and wasting a great deal of time for three or four weeks, I was obliged to relinquish the attempt. Soon afterwards I engaged in another branch of business (chandelier furniture), and took no more notice of it. About eighteen months ago I resumed the trinket trade, and then determined to think of the dolls' eyes; and about eight months since, I accidentally met with a poor fellow who had impoverished himself by drinking, and who was dying in a consumption, in a state of great want. I showed him ten sovereigns; and he said he would instruct me in the process. He was in such a state that he could not bear the effluvia of his own lamp; but though I was very conversant with the manual part of the business, and it related to things I was daily in the habit of seeing, I felt I could do nothing from his description. (I mention this to show how difficult it is to convey, by description, the mode of working.) He took me into his garret, where the poor fellow had economized to such a degree, that he actually used the entrails and fat of poultry from Leadenhall market to save oil (the price of the article having been lately so much reduced by competition at home). In an instant, before I had seen / him make three, I felt competent to make a gross; and the difference between his mode and that of my own workmen was so trifling, that I felt the utmost astonishment.

Question. You can now make dolls' eyes?

Answer. I can. As it was eighteen years ago that I received the order I have mentioned, and feeling doubtful of my own recollection, though very strong, and suspecting that it could [not] have been to the amount stated, I last night took the present very reduced price of that article (less than half now of what it was then), and calculating that every child in this country not using a doll till two years old, and throwing it aside at seven, and having a new one annually, I satisfied myself that the eyes alone would produce a circulation of a great many thousand pounds. I mention this merely to show the importance of trifles; and to assign one reason, amongst many, for my conviction, that nothing but personal communication can enable our manufactures to be transplanted.

301. In many instances it is exceedingly difficult to estimate beforehand the sale of an article, or the effects of a machine; a casc, however, occurred during a recent enquiry, which although not quite appropriate as an illustration of probable demand, is highly instructive as to the mode of conducting investigations of this nature. A committee of the House of Commons was appointed to enquire into the tolls proper to be placed on steam-carriages; a question, apparently, of difficult solution, and upon which widely different opinions had been formed, if we may judge by the very different rate of tolls imposed upon such carriages by different 'turnpike trusts'. The principles on which the committee conducted the enquiry were, that 'The only ground on which a fair claim to toll can be made on any public road, is to raise / a fund, which, with the strictest economy, shall be just sufficient – first, to repay the expense of its original formation; secondly, to maintain it in good and sufficient repair.' They first endeavoured to ascertain, from competent persons, the effect of the atmosphere alone in deteriorating a well-constructed road. The next step was, to determine the proportion in which the road was injured, by the effect of the horses' feet compared with that of the wheels. Mr Macneill, the superintendent, under Mr Telford, of the Holyhead roads, was examined, and proposed to estimate the relative injury, from the comparative quantities of iron worn off from the shoes of the horses, and from the tire of the wheels. From the data he possessed, respecting the consumption of iron for the tire of the wheels, and for the shoes of the horses, of one of the Birmingham day-coaches, he estimated the wear and tear of roads, arising from

the feet of the horses, to be three times as great as that arising from the wheels. Supposing repairs amounting to a hundred pounds to be required on a road travelled over by a fast coach at the rate of ten miles an hour, and the same amount of injury to occur on another road, used only by waggons, moving at the rate of three miles an hour, Mr Macneill divides the injuries in the following proportions: /

Injuries arising from	*Fast coach*	*Heavy waggon*
Atmospheric changes	20	20
Wheels	20	35·5
Horses' feet drawing	60	44·5
Total injury	100	100

Supposing it, therefore, to be ascertained that the wheels of steam carriages do no more injury to roads than other carriages of equal weight travelling with the same velocity, the committee now possessed the means of approximating to a just rate of toll for steam carriages.*

302. As connected with this subject, and as affording most valuable information upon points in which, previous to experiment, widely different opinions have been entertained; the following extract is inserted from Mr Telford's Report on the State of the Holyhead and Liverpool Roads. The instrument employed for the comparison was invented by Mr Macneill; and the road between London and Shrewsbury was selected for the place of experiment.

The general results, when a waggon weighing 21 cwt was used on different sorts of roads, are as follows: /

	lbs
1. On well-made pavement, the draught is	33
2. On a broken stone surface, or old flint road	65
3. On a gravel road	147
4. On a broken stone road, upon a rough pavement foundation	46
5. On a broken stone surface, upon a bottoming of concrete, formed of Parker's cement and gravel	46

The following statement relates to the force required to draw a coach weighing 18 cwt, exclusive of seven passengers, up roads of various inclinations:

* One of the results of these enquiries is, that every coach which travels from London to Birmingham distributes about eleven pounds of wrought-iron, along the line of road between those two places.

Inclination	Force required at six miles per hour	Force at eight miles per hour	Force at ten miles per hour
	lbs	lbs	lbs
1 in 20	268	296	318
1 in 26	213	219	225
1 in 30	165	196	200
1 in 40	160	166	172
1 in 600	111	120	128

303. In establishing a new manufactory, the time in which the goods produced can be brought to market and the returns be realized, should be thoroughly considered, as well as the time the new article will take to supersede those already in use. If it is destroyed in using, the new produce will be much more easily introduced. Steel pens readily took the place of quills; and a new form of pen would, if it possessed any advantage, as easily supersede the present one. A new lock, however secure, and however cheap, / would not so readily make its way. If less expensive than the old, it would be employed in new work: but old locks would rarely be removed to make way for it; and even if perfectly secure, its advance would be slow.

304. Another element in this question which should not be altogether omitted, is the opposition which the new manufacture may create by its real or apparent injury to other interests, and the probable effect of that opposition. This is not always foreseen; and when anticipated is often inaccurately estimated. On the first establishment of steam-boats from London to Margate, the proprietors of the coaches running on that line of road petitioned the House of Commons against them, as likely to lead to the ruin of the coach proprietors. It was, however, found that the fear was imaginary; and in a very few years, the number of coaches on that road was considerably increased, apparently through the very means which were thought to be adverse to it. The fear, which is now entertained, that steam-power and railroads may drive out of employment a large proportion of the horses at present in use, is probably not less unfounded. On some particular lines such an effect might be produced; but in all probability the number of horses employed in conveying goods and passengers to the great lines of railroad, would exceed that which is at present used. /

[250/1]

CHAPTER XXVI

ON A NEW SYSTEM OF MANUFACTURING

305. A most erroneous and unfortunate opinion prevails amongst workmen in many manufacturing countries, that their own interest and that of their employers are at variance. The consequences are – that valuable machinery is sometimes neglected, and even privately injured – that new improvements, introduced by the masters, do not receive a fair trial – and that the talents and observations of the workmen are not directed to the improvement of the processes in which they are employed. This error is, perhaps, most prevalent where the establishment of manufactories has been of recent origin, and where the number of persons employed in them is not very large: thus, in some of the Prussian provinces on the Rhine it prevails to a much greater extent than in Lancashire. Perhaps its diminished prevalence in our own manufacturing districts, arises partly from the superior information spread amongst the workmen; and partly from the frequent example of persons, who by good conduct and an attention to the interests of their employers for a series of years, have become foremen, or who have ultimately been admitted into advantageous partnerships. Convinced as I am, from my own observation, that the prosperity and / success of the master manufacturer is essential to the welfare of the workman, I am yet compelled to admit that this connection is, in many cases, too remote to be always understood by the latter: and whilst it is perfectly true that workmen, as a class, derive advantage from the prosperity of their employers, I do not think that each individual partakes of that advantage exactly in proportion to the extent to which he contributes towards it; nor do I perceive that the resulting advantage is as immediate as it might become under a different system.

306. It would be of great importance, if, in every large establishment the mode of payment could be so arranged, that every person employed should derive advantage from the success of the whole; and

that the profits of each individual should advance, as the factory itself produced profit, without the necessity of making any change in the wages. This is by no means easy to effect, particularly amongst that class whose daily labour procures for them their daily food. The system which has long been pursued in working the Cornish mines, although not exactly fulfilling these conditions, yet possesses advantages which make it worthy of attention, as having nearly approached towards them, and as tending to render fully effective the faculties of all who are engaged in it. I am the more strongly induced to place before the reader a short sketch of this system, because its similarity to that which I shall afterwards recommend for trial, will perhaps remove some objections to the latter, and may also furnish some valuable hints for conducting any experiment which might be undertaken. /

307. In the mines of Cornwall, almost the whole of the operations, both above and below ground, are contracted for. The manner of making the contract is nearly as follows. At the end of every two months, the work which it is proposed to carry on during the next period is marked out. It is of three kinds. 1. *Tutwork*, which consists in sinking shafts, driving levels, and making excavations: this is paid for by the fathom in depth, or in length, or by the cubic fathom. 2. *Tribute*, which is payment for raising and dressing the ore, by means of a certain part of its value when rendered merchantable. It is this mode of payment which produces such admirable effects. The miners, who are to be paid in proportion to the richness of the vein, and the quantity of metal extracted from it, naturally become quick-sighted in the discovery of ore, and in estimating its value; and it is their interest to avail themselves of every improvement that can bring it more cheaply to market. 3. *Dressing*. The 'Tributors', who dig and dress the ore, can seldom afford to dress the coarser parts of what they raise, at their contract price; this portion, therefore, is again let out to other persons, who agree to dress it at an advanced price.

The lots of ore to be dressed, and the works to be carried on, having been marked out some days before, and having been examined by the men, a kind of auction is held by the captains of the mine, in which each lot is put up, and bid for by different gangs of men. The work is then offered, at a price usually below that bid at the auction, to the lowest bidder, who rarely declines it at the rate proposed. The *tribute* is a certain sum out of every twenty shillings' / worth of ore, raised, and may vary from threepence to fourteen or fifteen shillings. The rate of earnings in tribute is very uncertain: if a vein, which was poor when taken, becomes rich, the men earn money rapidly; and instances have occurred in which each miner of a gang has gained a hundred pounds in the two months. These extraordinary cases, are, perhaps, of more advantage to the owners of the mine than even to the men; for whilst the skill and industry of the workmen are greatly stimulated, the owner himself always derives still greater advantage from the improvement of the vein.* This system has been introduced, by Mr Taylor, into the lead mines of Flintshire, into those at Skipton in Yorkshire, and into some of the copper mines of Cumberland; and it is desirable that it should become general, because no other mode of payment affords to the workmen a measure of success so directly proportioned to the industry, the integrity, and the talent, which they exert.

308. I shall now present the outline of a system which appears to me to be pregnant with the most important results, both to the class of workmen and to the country at large; and which, if acted upon, would, in my opinion, permanently raise the working classes, and greatly extend the manufacturing system.

The general principles on which the proposed system is founded, are

1. *That a considerable part of the wages received / by each person employed should depend on the profits made by the establishment; and,*

2. *That every person connected with it should derive more advantage from applying any improvement he might discover, to the factory in which he is employed, than he could by any other course.*

309. It would be difficult to prevail on the large capitalist to enter upon any system, which would change the division of the profits arising from the employment of his capital in setting skill and labour in action; any alteration, therefore, must be expected rather from the small capitalist, or from the higher class of workmen, who combine the two characters; and to these latter classes, whose welfare will be first affected, the change is most important. I shall therefore first point out the course to be pursued in making the experiment; and then, taking a particular branch of trade as an illustration, I shall examine the merits and defects of the proposed system as applied to it.

* For a detailed account of the method of working the Cornish mines, see a paper of Mr John Taylor's, *Transactions of the Geological Society*, vol. ii, p. 309. <John Taylor, 'On the economy of mines in Cornwall and Devon', *Transactions of the Geological Society*, Vol. 2 (1814), pp. 309–27.>

310. Let us suppose, in some large manufacturing town, ten or twelve of the most intelligent and skilful workmen to unite, whose characters for sobriety and steadiness are good, and are well known among their own class. Such persons will each possess some small portion of capital; and let them join with one or two others who have raised themselves into the class of small master manufacturers, and, therefore possess rather a larger portion of capital. Let these persons, after well considering the subject, agree to establish a manufactory of fire-irons and fenders; and let us suppose that each of the ten workmen can command forty pounds, and each of the small capitalists possesses two hundred pounds: thus they have / a capital of £800 with which to commence business; and, for the sake of simplifying, let us further suppose the labour of each of these twelve persons to be worth two pounds a week. One portion of their capital will be expended in procuring the tools necessary for their trade, which we shall take at £400, and this must be considered as their fixed capital. The remaining £400 must be employed as circulating capital, in purchasing the iron with which their articles are made, in paying the rent of their workshops, and in supporting themselves and their families until some portion of it is replaced by the sale of the goods produced.

311. Now the first question to be settled is, what proportion of the profit should be allowed for the use of capital, and what for skill and labour? It does not seem possible to decide this question by any abstract reasoning: if the capital supplied by each partner is equal, all difficulty will be removed; if otherwise, the proportion must be left to find its level, and will be discovered by experience; and it is probable that it will not fluctuate much. Let us suppose it to be agreed that the capital of £800 shall receive the wages of one workman. At the end of each week every workman is to receive one pound as wages, and one pound is to be divided amongst the owners of the capital. After a few weeks the returns will begin to come in; and they will soon become nearly uniform. Accurate accounts should be kept of every expense and of all the sales; and at the end of each week the profit should be divided. A certain portion should be laid aside as a reserved fund, another portion for repair of the tools, and the / remainder being divided into thirteen parts, one of these parts would be divided amongst the capitalists and one belong to each workman. Thus each man would, in ordinary circumstances, make up his usual wages of two pounds weekly. If the factory went on prosperously, the wages of

the men would increase; if the sales fell off they would be diminished. It is important that every person employed in the establishment, whatever might be the amount paid for his services, whether he act as labourer or porter, as the clerk who keeps the accounts, or as book-keeper employed for a few hours once a week to superintend them, should receive one half of what his service is worth in fixed salary, the other part varying with the success of the undertaking.

312. In such a factory, of course, division of labour would be introduced; some of the workmen would be constantly employed in forging the fire-irons, others in polishing them, others in piercing and forming the fenders. It would be essential that the time occupied in each process, and also its expense, should be well ascertained; information which would soon be obtained very precisely. Now, if a workman should find a mode of shortening any of the processes, he would confer a benefit on the whole party, even if they received but a small part of the resulting profit. For the promotion of such discoveries, it would be desirable that those who make them should either receive some reward, to be determined after a sufficient trial by a committee assembling periodically; or if they be of high importance, that the discoverer should receive one-half, or two-thirds, of the profit resulting from them during the next / year, or some other determinate period, as might be found expedient. As the advantages of such improvements would be clear gain to the factory, it is obvious that such a share might be allowed to the inventor, that it would be for his interest rather to give the benefit of them to his partners, than to dispose of them in any other way.

313. The result of such arrangements in a factory would be,

1. That every person engaged in it would have a *direct* interest in its prosperity; since the effect of any success, or falling off, would almost immediately produce a corresponding change in his own weekly receipts.

2. Every person concerned in the factory would have an immediate interest in preventing any waste or mismanagement in all the departments.

3. The talents of all connected with it would be strongly directed to its improvement in every department.

4. None but workmen of high character and qualifications could obtain admission into such establishments; because when any additional hands were required, it would be the common interest of all to admit only the most respectable and skilful; and it would be far

less easy to impose upon a dozen workmen than upon the single proprietor of a factory.

5. When any circumstance produced a glut in the market, more skill would be directed to diminishing the cost of production; and a portion of the time of the men might then be occupied in repairing and improving their tools, for which a reserved fund would pay, thus checking present, and at the same time facilitating future production. /

6. Another advantage, of no small importance, would be the total removal of all real or imaginary causes for combinations. The workmen and the capitalist would so shade into each other – would so *evidently* have a common interest, and their difficulties and distresses would be mutually so well understood that, instead of combining to oppress one another, the only combination which could exist would be a most powerful union *between* both parties to overcome their common difficulties.

314. One of the difficulties attending such a system is, that capitalists would at first fear to embark in it, imagining that the workmen would receive too large a share of the profits: and it is quite true that the workmen would have a larger share than at present: but, at the same time, it is presumed the effect of the whole system would be, that the total profits of the establishment being much increased, the smaller proportion allowed to capital under this system would yet be greater in actual amount, than that which results to it from the larger share in the system now existing.

315. It is possible that the present laws relating to partnerships might interfere with factories so conducted. If this interference could not be obviated by confining their purchases under the proposed system to ready money, it would be desirable to consider what changes in the law would be necessary to its existence: and this furnishes another reason for entering into the question of limited partnerships.

316. A difficulty would occur also in discharging workmen who behaved ill, or who were not competent / to their work; this would arise from their having a certain interest in the reserved fund, and, perhaps, from their possessing a certain portion of the capital employed; but without entering into detail, it may be observed, that such cases might be determined on by meetings of the whole establishment; and that if the policy of the laws favoured such establishments, it would scarcely be more difficult to enforce just regulations, than it now is to enforce some which are unjust, by means of combinations either amongst the masters or the men.

317. Some approach to this system is already practised in several trades: the mode of conducting the Cornish mines has already been alluded to; the payment to the crew of whaling ships is governed by this principle; the profits arising from fishing with nets on the south coast of England are thus divided: one-half the produce belongs to the owner of the boat and net; the other half is divided in equal portions between the persons using it, who are also bound to assist in repairing the net when injured. /

[260/1]

CHAPTER XXVII

ON CONTRIVING MACHINERY

318. The power of inventing mechanical contrivances, and of combining machinery, does not appear, if we may judge from the frequency of its occurrence, to be a difficult or a rare gift. Of the vast multitude of inventions which have been produced almost daily for a series of years, a large part has failed from the imperfect nature of the first trials; whilst a still larger portion, which had escaped thc mechanical difficulties, failed only because the economy of their operations was not sufficiently attended to.

The commissioners appointed to examine into the methods proposed for preventing the forgery of bank-notes, state in their report, that out of one hundred and seventy-eight projects communicated to the bank and to the commissioners, there were only twelve of superior skill, and nine which it was necessary more particularly to examine.

319. It is however a curious circumstance, that although the power of combining machinery is so common, yet the more beautiful combinations are exceedingly rare. Those which command our admiration equally by the perfection of their effects and the simplicity of their means, are found only amongst the happiest productions of genius. /

To produce movements even of a complicated kind is not difficult. There exist a great multitude of known contrivances for all the more usual purposes, and if the exertion of moderate power is the end of the mechanism to be contrived, it is possible to construct the whole machine upon paper, and to judge of the proper strength to be given to each part as well as to the framework which supports it, and also of its ultimate effect, long before a single part of it has been executed. In fact, all the contrivance, and all the improvements, ought first to be represented in the drawings.

320. On the other hand, there are effects dependent upon physical or chemical properties for the determination of which no drawings will

be of any use. These are the legitimate objects of direct trial. For example; if the ultimate result of an engine is to be that it shall impress letters on a copperplate by means of steel punches forced into it, all the mechanism by which the punches and the copper are to be moved at stated intervals, and brought into contact, is within the province of drawing, and the machinery may be arranged entirely upon paper. But a doubt may reasonably spring up, whether the bur that will be raised round the letter, which has been already punched upon the copper, may not interfere with the proper action of the punch for the letter which is to be punched next adjacent to it. It may also be feared that the effect of punching the second letter, if it be sufficiently near to the first, may distort the form of that first figure. If neither of these evils should arise, still the bur produced by the punching might be expected to interfere with the / goodness of the impression produced by the copperplate; and the plate itself, after having all but its edge covered with figures, might change its form, from the unequal condensation which it must suffer in this process, so as to render it very difficult to take impressions from it at all. It is impossible by any drawings to solve difficulties such as these, experiment alone can determine their effect. Such experiments having been made, it is found that if the sides of the steel punch are nearly at right angles to the face of the letter, the bur produced is very inconsiderable; that at the depth which is sufficient for copperplate printing, no distortion of the adjacent letters takes place, although those letters are placed very close to each other; that the small bur which arises may easily be scraped off; and that the copperplate is not distorted by the condensation of the metal in punching, but is perfectly fit to print from, after it has undergone that process.

321. The next stage in the progress of an invention, after the drawings are finished and the preliminary experiments have been made, if any such should be requisite, is the execution of the machine itself. It can never be too strongly impressed upon the minds of those who are devising new machines, that to make the most perfect drawings of every part tends essentially both to the success of the trial, and to economy in arriving at the result. The actual execution from working drawings is comparatively an easy task; provided always that good tools are employed, and that methods of working are adopted, in which the perfection of the part constructed depends / less on the personal skill of the workman, than upon the certainty of the method employed.

322. The causes of failure in this stage most frequently derive their origin from errors in the preceding one; and it is sufficient merely to indicate a few of their sources. They frequently arise from having neglected to take into consideration that metals are not perfectly rigid but elastic. A steel cylinder of small diameter must not be regarded as an inflexible rod; but in order to ensure its perfect action as an axis, it must be supported at proper intervals.

Again, the strength and stiffness of the framing which supports the mechanism must be carefully attended to. It should always be recollected, that the addition of superfluous matter to the immovable parts of a machine produces no additional momentum, and therefore is not accompanied with the same evil that arises when the moving parts are increased in weight. The stiffness of the framing in a machine produces an important advantage. If the bearings of the axis (those places at which they are supported) are once placed in a straight line, they will remain so, if the framing be immovable; whereas if the framework changes its form, though ever so slightly, considerable friction is immediately produced. This effect is so well understood in the districts where spinning factories are numerous, that, in estimating the expense of working a new factory, it is allowed that five per cent on the power of the steam-engine will be saved if the building is fireproof: for the greater strength and rigidity of a fireproof building prevents the movement of the long shafts or axes which drive the machinery, from being / impeded by the friction that would arise from the slightest deviation in any of the bearings.

323. In conducting experiments upon machinery, it is quite a mistake to suppose that any imperfect mechanical work is good enough for such a purpose. If the experiment is worth making, it ought to be tried with all the advantages of which the state of mechanical art admits; for an imperfect trial may cause an idea to be given up, which better workmanship might have proved to be practicable. On the other hand, when once the efficiency of a contrivance has been established, with good workmanship, it will be easy afterwards to ascertain the degree of perfection which will suffice for its due action.

324. It is partly owing to *the imperfection of the original trials*, and partly to the gradual improvements in the art of making machinery, that many inventions which have been tried, and given up in one state of art, have at another period been eminently successful. The

idea of printing by means of moveable types had probably suggested itself to the imagination of many persons conversant with impressions taken either from blocks or seals. We find amongst the instruments discovered in the remains of Pompeii and Herculaneum, stamps for words formed out of one piece of metal, and including several letters. The idea of separating these letters, and of recombining them into other words, for the purpose of stamping a book, could scarcely have failed to occur to many: but it would almost certainly have been rejected by those best acquainted with the mechanical arts of that time; for the workmen of those days must have instantly perceived the / impossibility of producing many thousand pieces of wood or metal, fitting so perfectly and ranging so uniformly, as the types or blocks of wood now used in the art of printing.

The principle of the press which bears the name of Bramah, was known about a century and a half before the machine, to which it gave rise, existed; but the imperfect state of mechanical art in the time of the discoverer, would have effectually deterred him, if the application of it had occurred to his mind, from attempting to employ it in practice as an instrument for exerting force.

These considerations prove the propriety of repeating, at the termination of intervals during which the art of making machinery has received any great improvement, the trails of methods which, although founded upon just principles, had previously failed.

325. When the drawings of a machine have been properly made, and the parts have been well executed, and even when the work it produces possesses all the qualities which were anticipated, still the invention may fail; that is, *it may fail of being brought into general practice.* This will most frequently arise from the circumstance of its producing its work at a greater expense than that at which it can be made by other methods.

326. Whenever the new, or improved machine, is intended to become the basis of a manufacture, it is essentially requisite that the *whole* expense attending its operations should be fully considered before its construction is undertaken. It is almost always very difficult to make this estimate of the expense: the more complicated the mechanism, the less easy is / the task; and in cases of great complexity and extent of machinery it is almost impossible. It has been estimated roughly, that the first individual of any newly invented machine, will cost about five times as much as the construction of the second, an estimate which is, perhaps, sufficiently near the truth. If the second

machine is to be precisely like the first, the same drawings, and the same patterns will answer for it; but if, as usually happens, some improvements have been suggested by the experience of the first, these must be more or less altered. When, however, two or three machines have been completed, and many more are wanted, they can usually be produced at much less than one-fifth of the expense of the original invention.

327. The arts of contriving, of drawing, and of executing, do not usually reside in their greatest perfection in one individual; and in this, as in other arts, *the division of labour* must be applied. The best advice which can be offered to a projector of any mechanical invention, is to employ a respectable draughtsman; who, if he has had a large experience in his profession, will assist in finding out whether the contrivance is new, and can then make working drawings of it. The first step, however, the ascertaining whether the contrivance has the merit of novelty, is most important; for it is a maxim equally just in all the arts, and in every science, that *the man who aspires to fortune or to fame by new discoveries, must be content to examine with care the knowledge of his contemporaries, or to exhaust his efforts in inventing again, what he will most probably find has been better executed before.* /

328. This, nevertheless, is a subject upon which even ingenious men are often singularly negligent. There is, perhaps, no trade or profession existing in which there is so much quackery, so much ignorance of the scientific principles, and of the history of their own art, with respect to its resources and extent, as are to be met with amongst mechanical projectors. The self-constituted engineer, dazzled with the beauty of some, perhaps, really original contrivance, assumes his new profession with as little suspicion that previous instruction, that thought and painful labour, are necessary to its successful exercise, as does the statesman or the senator. Much of this false confidence arises from the improper estimate which is entertained of the difficulty of invention in mechanics. It is, therefore, of great importance to the individuals and to the families of those who are too often led away from more suitable pursuits, the dupes of their own ingenuity and of the popular voice, to convince both them and the public that the power of making new mechanical combinations is a possession common to a multitude of minds, and that the talents which it requires are by no means of the highest order. It is still more important that they should be impressed with

the conviction that the great merit, and the great success of those who have attained to eminence in such matters, was almost entirely due to the unremitted perseverance with which they concentrated upon their successful inventions the skill and knowledge which years of study had matured. /

[268/9]

CHAPTER XXVIII

PROPER CIRCUMSTANCES FOR THE APPLICATION OF MACHINERY

329. The first object of machinery, the chief cause of its extensive utility, is the perfection and the cheap production of the articles which it is intended to make. Whenever it is required to produce a great multitude of things, all of exactly the same kind, the proper time has arrived for the construction of tools or machines by which they may be manufactured. If only a few pairs of cotton stockings should be required, it would be an absurd waste of time, and of capital, to construct a stocking-frame to weave them, when, for a few pence, four steel wires can be procured by which they may be knit. If, on the other hand, many thousand pairs were wanted, the time employed, and the expense incurred in constructing a stocking-frame, would be more than repaid by the saving of time in making that large number of stockings. The same principle is applicable to the copying of letters: if three or four copies only are required, the pen and the human hand furnish the cheapest means of obtaining them; if hundreds are called for, lithography may be brought to our assistance; but if hundreds of thousands are wanted, the machinery of a printing establishment supplies the most economical method of accomplishing the object. /

330. There are, however, many cases in which machines or tools must be made, in which economical production is not the most important object. Whenever it is required to produce a few articles – parts of machinery, for instance, which must be executed with the most rigid accuracy or be perfectly alike – it is nearly impossible to fulfil this condition, even with the aid of the most skilful hands: and it becomes necessary to make tools expressly for the purpose, although those tools should, as frequently happens, cost more in constructing than the things they are destined to make.

331. Another instance of the just application of machinery, even at an

increased expense, arises where the shortness of time in which the article is produced, has an important influence on its value. In the publication of our daily newspapers, it frequently happens that the debates in the Houses of Parliament are carried on to three and four o'clock in the morning, that is, to within a very few hours of the time for the publication of the paper. The speeches must be taken down by reporters, conveyed by them to the establishment of the newspaper, perhaps at the distance of one or two miles, transcribed by them in the office, set up by the compositor, the press corrected, and the paper be printed off and distributed, before the public can read them. Some of these journals have a circulation of from five to ten thousand daily. Supposing four thousand to be wanted, and that they could be printed only at the rate of five hundred per hour upon one side of the paper, (which was the greatest number two journeymen and a boy could take off by the old hand presses), sixteen / hours would be required for printing the complete edition; and the news conveyed to the purchasers of the latest portion of the impression, would be out of date before they could receive it. To obviate this difficulty, it was often necessary to set up the paper in duplicate, and sometimes, when late, in triplicate: but the improvements in the printing machines have been so great, that four thousand copies are now printed on one side in an hour.

332. The establishment of 'The Times' newspaper is an example, on a large scale, of a manufactory in which the division of labour, both mental and bodily, is admirably illustrated, and in which also the effect of domestic economy is well exemplified. It is scarcely imagined by the thousands who read that paper in various quarters of the globe, what a scene of organized activity the factory presents during the whole night, or what a quantity of talent and mechanical skill is put in action for their amusement and information.* Nearly a

* The author of these pages, with one of his friends, was recently induced to visit this most interesting establishment, after midnight, during the progress of a very important debate. The place was illuminated with gas, and was light as the day: there was neither noise nor bustle; and the visitors were received with such calm and polite attention, that they did not, until afterwards, become sensible of the inconvenience which such intruders, at a moment of the greatest pressure, must occasion, nor reflect that the tranquillity which they admired, was the result of intense and regulated occupation. But the effect of such checks in the current of business will appear on recollecting that, as four thousand newspapers are printed off on one side within the hour, every *minute* is attended with a loss of sixty-six impressions. The quarter of an hour, therefore, which the stranger may think it not unreasonable to claim for the gratification of his curiosity (and to him this time is but a moment), may cause a failure in the delivery of a

hundred persons are / employed in this establishment; and, during the session of Parliament, at least twelve reporters are constantly attending the Houses of Commons and Lords; each in his turn retiring, after about an hour's work, to translate into ordinary writing, the speech he has just heard and noted in shorthand. In the meantime fifty compositors are constantly at work, some of whom have already set up the beginning, whilst others are committing to type the yet undried manuscript of the continuation of a speech, whose middle portion is travelling to the office in the pocket of the hasty reporter, and whose eloquent conclusion is, perhaps, at that very moment, making the walls of St Stephen's vibrate with the applause of its hearers. These congregated types, as fast as they are composed, are passed in portions to other hands; till at last the scattered fragments of the debate, forming, / when united with the ordinary matter, eight-and-forty columns, reappear in regular order on the platform of the printing-press. The hand of man is now too slow for the demands of his curiosity, but the power of steam comes to his assistance. Ink is rapidly supplied to the moving types, by the most perfect mechanism; four attendants incessantly introduce the edges of large sheets of white paper to the junction of two great rollers, which seem to devour them with unsated appetite; other rollers convey them to the type already inked, and having brought them into rapid and successive contact, redeliver them to four other assistants, completely printed by the almost momentary touch. Thus, in one hour, four thousand sheets of paper are printed on one side; and an impression of twelve thousand copies, from above three hundred thousand moveable pieces of metal, is produced for the public in six hours.

333. The effect of machinery in printing other periodical publications,

---

thousand copies, and disappoint a proportionate number of expectant readers, in some of our distant towns, to which the morning papers are despatched by the earliest and most rapid conveyances of each day.

This note is inserted with the further and more general purpose of calling the attention of those, especially foreigners, who are desirous of inspecting our larger manufactories, to the chief cause of the difficulty which frequently attends their introduction. When the establishment is very extensive, and its departments skilfully arranged, the exclusion of visitors arises, not from any illiberal jealousy, nor, generally, from any desire of concealment, which would, in most cases, be absurd; but from the substantial inconvenience and loss of time, throughout an entire series of well-combined operations, which must be occasioned even by short and casual interruptions.

and of due economy in distributing them, is so important for the interests of knowledge, that it is worth examining by what means it is possible to produce them at the small price at which they are sold. 'Chambers' Journal', which is published at Edinburgh, and sold at three halfpence a number, will furnish an example. Soon after its commencement in 1832, the sale in Scotland reached 30,000, and in order to supply the demand in London it was reprinted; but on account of the expense of 'composition' it was found that this plan would not produce any profit, and the London edition was about to be given up, when it occurred to the proprietor / to stereotype it at Edinburgh, and cast two copies of the plates. This is now done about three weeks before the day of publication – one set of plates being sent up to London by the mail, an impression is printed off by steam: the London agent has then time to send packages by the cheapest conveyances to several of the large towns, and other copies go through the booksellers' parcels to all the smaller towns. Thus a great saving is effected in the outlay of capital, and 20,000 copies are conveyed from London, as a centre, to all parts of England, whilst there is no difficulty in completing imperfect sets, nor any waste from printing more than the public demand.

334. The conveyance of letters is another case, in which the importance of saving time would allow of great expense in any new machinery for its accomplishment. There is a natural limit to the speed of horses, which even the greatest improvements in the breed, aided by an increased perfection in our roads, can never surpass; and from which, perhaps, we are at present not very remote. When we reflect upon the great expense of time and money which the last refinements of a theory or an art usually require, it is not unreasonable to suppose that the period has arrived in which the substitution of machinery for such purposes ought to be tried.

335. The post bag despatched every evening by the mail to one of our largest cities, Bristol, usually weighs less than a hundred pounds. Now, the first reflection which naturally presents itself is, that, in order to transport these letters a hundred and twenty miles, a coach and apparatus, weighing above thirty / hundredweight, are put in motion, and also conveyed over the same space.*

It is obvious that, amongst the conditions of machinery for

* It is true that the transport of letters is not the only object which this apparatus answers; but the transport of passengers, which is a secondary object, does in fact put a limit to the velocity of that of the letters, which is the primary one.

accomplishing such an object, it would be desirable to reduce the weight of matter to be conveyed along with the letters: it would also be desirable to reduce the velocity of the animal power employed; because the faster a horse is driven, the less weight he can draw. Amongst the variety of contrivances which might be imagined for this purpose, we will mention one, which, although by no means free from objections, fulfils some of the prescribed conditions; and it is not a purely theoretical speculation, since some few experiments have been made upon it, though on an extremely limited scale.

336. Let us imagine a series of high pillars erected at frequent intervals, perhaps every hundred feet, and as nearly as possible in a straight line between two post towns. An iron or steel wire must be stretched over proper supports, fixed on each of these pillars, and terminating at the end of every three or five miles, as may be found expedient, in a very strong support, by which it may be stretched. At each of these latter points a man ought to reside in a small station-house. A narrow cylindrical tin case, to contain the letters, might be suspended by two wheels rolling upon this wire; the cases being so constructed as to enable the wheels to pass unimpeded / by the fixed supports of the wire. An endless wire of much smaller size must pass over two drums, one at each end of the station. This wire should be supported on rollers, fixed to the supports of the great wire, and at a short distance below it. There would thus be two branches of the smaller wire always accompanying the larger one; and the attendant at either station, by turning the drum, might cause them to move with great velocity in opposite directions. In order to convey the cylinder which contains the letters, it would only be necessary to attach it by a string, or by a catch, to either of the branches of the endless wire. Thus it would be conveyed speedily to the next station, where it would be removed by the attendant to the commencement of the next wire, and so forwarded. It is unnecessary to enter into the details which this, or any similar plan, would require. The difficulties are obvious; but if these could be overcome, it would present many advantages besides velocity; for if an attendant resided at each station, the additional expense of having two or three deliveries of letters every day, and even of sending expresses at any moment, would be comparatively trifling; nor is it impossible that the stretched wire might itself be available for a species of telegraphic communication yet more rapid.

Perhaps if the steeples of churches, properly selected, were made use of, connecting them by a few intermediate stations with some great central building, as, for instance, with the top of St Paul's; and if a similar apparatus were placed on the top of each steeple, with a man to work it during the day, / it might be possible to diminish the expense of the two-penny post, and make deliveries every half hour over the greater part of the metropolis.

337. The power of steam, however, bids fair almost to rival the velocity of these contrivances; and the fitness of its application to the purposes of conveyance, particularly where great rapidity is required, begins now to be generally admitted. The following extract from the Report of the Committee of the House of Commons on steam-carriages, explains clearly its various advantages:

Perhaps one of the principal advantages resulting from the use of steam, will be, that it may be employed as cheaply at a quick as at a slow rate; 'this is one of the advantages over horse labour, which becomes more and more expensive as the speed is increased. There is every reason to expect, that in the end the rate of travelling by steam will be much quicker than the utmost speed of travelling by horses; in short, the safety to travellers will become the limit to speed.' In horse-draught the opposite result takes place; 'in all cases horses lose power of draught in a much greater proportion than they gain speed, and hence the work they do becomes more expensive as they go quicker.'

Without increase of cost, then, we shall obtain a power which will insure a rapidity of internal communication far beyond the utmost speed of horses in draught; and although the performance of these carriages may not have hitherto attained this point, when once it has been established, that at equal speed we can use steam more cheaply in draught than horses, we may fairly anticipate that every day's increased experience in the management of the engines, will induce greater skill, greater confidence, and greater speed.

The cheapness of the conveyance will probably be, for some time, a secondary consideration. If, at present, it / can be used as cheaply as horse power, the competition with the former modes of conveyance will first take place as to speed. When once the superiority of steam-carriages shall have been fully established, competition will induce economy in the cost of working them. The evidence, however, of Mr Macneill, showing the greater efficiency, with diminished expenditure of fuel, by locomotive engines on railways, convinces the committee, that experience will soon teach a better construction of the engines, and a less costly mode of generating the requisite supply of steam.

Nor are the advantages of steam-power confined to the greater velocity attained, or to its greater cheapness than horse-draught. In the latter, danger is increased, in as large a proportion as expense, by greater speed. In steam-power, on the contrary, 'there is no danger of being run away with, and that of being overturned is greatly diminished. It is difficult to control four such horses as can draw a heavy carriage ten miles per hour, in case they are frightened, or choose to run away; and for quick travelling they must be kept in that state of courage, that they are always inclined for running away, particularly down hills, and at sharp turns of the road. In steam, however, there is little corresponding danger, being perfectly controllable, and capable of exerting its power in reverse in going down hills.' Every witness examined has given the fullest and most satisfactory evidence of the perfect control which the conductor has over the movement of the carriage. With the slightest exertion it can be stopped or turned, under circumstances where horses would be totally unmanageable.

338. Another instance may be mentioned in which the object to be obtained is so important, that although it might be rarely wanted, yet machinery for that purpose would justify considerable expense. A vessel to contain men, and to be navigated at some / distance below the surface of the sea, would, in many circumstances, be almost invaluable. Such a vessel, evidently, could not be propelled by any engine requiring the aid of fire. If, however, by condensing air into a liquid, and carrying it in that state, a propelling power could be procured sufficient for moving the vessel through a considerable space, the expense would scarcely render its occasional employment impossible.*

339. *Slide of Alpnach.* Amongst the forests which flank many of the lofty mountains of Switzerland, some of the finest timber is found in positions almost inaccessible. The expense of roads, even if it were possible to make them in such situations, would prevent the inhabitants from deriving any advantages from these almost inexhaustible supplies. Placed by nature at a considerable elevation above the spot at which they can be made use of, they are precisely in fit circumstances for the application of machinery to their removal; and the inhabitants avail themselves of the force of gravity to relieve them from some portion of this labour. The inclined planes which they have established in various forests, by which the timber has been sent down to the water courses, have excited the admiration of every traveller; and in addition to the merit of simplicity, the construction

* A proposal for such a vessel, and description of its construction, by the author of this volume, may be found in the *Encyclopaedia Metropolitana*, Art. Diving Bell. <(1826), *Works of Babbage*, Vol. 4.>

these slides requires scarcely anything beyond the material which grows upon the spot.

Of all these specimens of carpentry, the Slide of Alpnach was the most considerable, from its great / length, and from the almost inaccessible position from which it descended. The following account of it is taken from Gilbert's Annalen, 1819, which is translated in the second volume of Brewster's Journal:

For many centuries, the rugged flanks and the deep gorges of Mount Pilatus were covered with impenetrable forests; which were permitted to grow and to perish, without being of the least utility to man, till a foreigner, who had been conducted into their wild recesses in the pursuit of the chamois, directed the attention of several Swiss gentlemen to the extent and superiority of the timber. The most skilful individuals, however, considered it quite impracticable to avail themselves of such inaccessible stores. It was not till the end of 1816, that M. Rupp, and three Swiss gentlemen, entertaining more sanguine hopes, purchased a certain extent of the forests, and began the construction of the slide, which was completed in the spring of 1818.

The Slide of Alpnach is formed entirely of about 25,000 large pine trees, deprived of their bark, and united together in a very ingenious manner, without the aid of iron. It occupied about 160 workmen during eighteen months, and cost nearly 100,000 francs, or £4,250. It is about three leagues, or 44,000 English feet long, and terminates in the Lake of Lucerne. It has the form of a trough, about six feet broad, and from three to six feet deep. Its bottom is formed of three trees, the middle one of which has a groove cut out in the direction of its length, for receiving small rills of water, which are conducted into it from various places, for the purpose of diminishing the friction. The whole of the / slide is sustained by about 2,000 supports; and in many places it is attached, in a very ingenious manner, to the rugged precipices of granite.

The direction of the slide is sometimes straight, and sometimes zig-zag, with an inclination of from 10° to 18°. It is often carried along the sides of hills and the flanks of precipitous rocks, and sometimes passes over their summits. Occasionally it goes under ground, and at other times it is conducted over the deep gorges by scaffoldings 120 feet in height.

The boldness which characterizes this work, the sagacity and skill displayed in all its arrangements, have excited the wonder of every person who has seen it. Before any step could be taken in its erection, it was necessary to cut several thousand trees to obtain a passage through the impenetrable thickets. All these difficulties, however, were surmounted, and the engineer had at last the satisfaction of seeing the trees descend from the mountain with the rapidity of lightning. The larger pines, which were about a hundred feet long, and ten inches thick at their smaller extremity, ran through the space of *three leagues*, or nearly *nine miles*, *in two minutes and a half*, and during their descent, they appeared to be only a few feet in length. The arrangements for this part of the operation were extremely simple. From the lower end of the slide to the upper end, where the trees were introduced, workmen were posted at regular distances, and as soon as everything was ready, the workman at the lower end of the slide cried out to the one above him, '*Lachez*' (let go). The cry / was repeated from one to another, and reached the top of the slide in *three* minutes. The workmen at the top of the slide then cried out to the one below him, '*Il vient*' (it comes), and the tree was instantly launched down the slide, preceded by the cry which was repeated from post to post. As soon as the tree had reached the bottom, and plunged into the lake, the cry of *lachez* was repeated as before, and a new tree was launched in a similar manner. By these means a tree descended every five or six minutes, provided no accident happened to the slide, which sometimes took place, but which was instantly repaired when it did.

In order to show the enormous force which the trees acquired from the great velocity of their descent, M. Rupp made arrangements for causing some of the trees to spring from the slide. They penetrated by their thickest extremities no less than from eighteen to twenty-four feet into the earth; and one of the trees having by accident struck against another, it instantly cleft it through its whole length, as if it had been struck by lightning.

After the trees had descended the slide, they were collected into rafts upon the lake, and conducted to Lucerne. From thence they descended the Reuss, then the Aar to near Brugg, afterwards to Waldshut by the Rhine, then to Basle, and even to the sea when it was necessary.

It is to be regretted that this magnificent structure no longer exists, and that scarcely a trace of it is to be seen upon the flanks of Mount Pilatus. Political circumstances having taken away the principal source of demand for the timber, and no other / market having been found, the operation of cutting and transporting the trees necessarily ceased.*

Professor Playfair, who visited this singular work, states, that six minutes was the usual time occupied in the descent of a tree; but that in wet weather, it reached the lake in three minutes. /

* The mines of Bolanos in Mexico are supplied with timber from the adjacent mountains by a slide similar to that of Alpnach. It was constructed by M. Floresi, a gentleman well acquainted with Switzerland.

CHAPTER XXIX

ON THE DURATION OF MACHINERY

340. The time during which a machine will continue to perform its work effectually, will depend chiefly upon the perfection with which it was originally constructed – upon the care taken to keep it in proper repair, particularly to correct every shake or looseness in the axes – and upon the smallness of the mass and of the velocity of its moving parts. Everything approaching to a blow, all sudden change of direction, is injurious. Engines for producing power, such as windmills, water-mills, and steam-engines, usually last a long time.*

341. Many of the improvements which have taken place in steam-engines, have arisen from an improved construction of the boiler or the fireplace. The following table of the work done by steam-engines in Cornwall, whilst it proves the importance of constantly measuring the effects of machinery, shows also the gradual advance which has been made in the art of constructing and managing those engines. /

A table of the duty performed by steam-engines in Cornwall, showing the average of the whole for each year, and also the average duty of the best engine in each monthly report

Years	*Approximate number of engines reported*	*Average duty of the whole*	*Average duty of the best engines*
1813	24	19,456,000	26,400,000
1814	29	20,534,232	32,000,000
1815	35	20,526,160	28,700,000
1816	32	22,907,110	32,400,000
1817	31	26,502,259	41,600,000
1818	32	25,433,783	39,300,000

* The return which ought to be produced by a fixed steam-engine employed as a moving power, is frequently estimated at ten per cent on its cost.

Years	*Approximate number of engines reported*	*Average duty of the whole*	*Average duty of the best engines*
1819	37	26,252,620	40,000,000
1820	37	28,736,398	41,300,000
1821	39	28,223,382	42,800,000
1822	45	28,887,216	42,500,000
1823	45	28,156,162	42,122,000
1824	45	28,326,140	43,500,000
1825	50	32,000,741	45,400,000
1826	48	30,486,630	45,200,000
1827	47	32,100,000	59,700,000
1828	54	37,100,000	76,763,000
1829	52	41,220,000	76,234,307
1830	55	43,350,000	75,885,519
1831	55*	44,700,000	74,911,365
1832	60	44,400,000	79,294,114
1833	58	46,000,000	83,306,092

342. The advantage arising from registering the duty done by steam-engines in Cornwall has been so great that the proprietors of one of the largest mines, on which there are several engines, find it good economy to employ a man to measure the duty they perform every day. This daily report is fixed up at a particular hour, and the engine-men are always in waiting, anxious to know the state of their engines. As the general reports are made monthly, if accident / should cause a partial stoppage in the flue of any of the boilers, it might without this daily check continue two or three weeks before it could be discovered by a falling off of the duty of the engine. In several of the mines a certain amount of duty is assigned to each engine; and if it does more, the proprietors give a premium to the engineers according to its amount. This is called million money, and is a great stimulus to economy in working the engine.

343. Machinery for producing any commodity in great demand, seldom actually wears out; new improvements, by which the same operations can be executed either more quickly or better, generally superseding it long before that period arrives: indeed, to make such an improved machine profitable, it is usually reckoned that in five years it ought to have paid itself, and in ten to be superseded by a better.

* These fifty-five engines consumed, on the average for the year 1831, 81,867 bushels of coals monthly, being 1488 bushels for each engine.

'A cotton manufacturer,' says one of the witnesses before a Committee of the House of Commons, 'who left Manchester seven years ago, would be driven out of the market by the men who are now living in it, provided his knowledge had not kept pace with those who have been, during that time, constantly profiting by the progressive improvements that have taken place in that period.'

344. The effect of improvements in machinery, seems incidentally to increase production, through a cause which may be thus explained. A manufacturer making the usual profit upon his capital, invested in looms or other machines in perfect condition, the market price of making each of which is a hundred pounds, invents some improvement. But this is of such a nature, that it cannot be adapted to his present engines. He finds upon calculation, / that at the rate at which he can dispose of his manufactured produce, each new engine would repay the cost of its making, together with the ordinary profit of capital, in three years: he also concludes from his experience of the trade, that the improvement he is about to make, will not be generally adopted by other manufacturers before that time. On these considerations, it is clearly his interest to sell his present engines, even at half-price, and construct new ones on the improved principle. But the purchaser who gives only fifty pounds for the old engines, has not so large a fixed capital invested in his factory, as the person from whom he purchased them; and as he produces the same quantity of the manufactured article, his profits will be larger. Hence, the price of the commodity will fall, not only in consequence of the cheaper production by the new machines, but also by the more profitable working of the old, thus purchased at a reduced price. This change, however, can be only transient; for a time will arrive when the old machinery, although in good repair, must become worthless. The improvement which took place not long ago in frames for making patent-net was so great, that a machine, in good repair, which had cost £1200, sold a few years after for £60. During the great speculations in that trade, the improvements succeeded each other so rapidly, that machines which had never been finished were abandoned in the hands of their makers, because new improvements had superseded their utility.

345. The durability of watches, when well made, is very remarkable. One was produced, in *going order*, before a committee of the House of / Commons to enquire into the watch trade, which was made in the year 1660; and there are many of ancient date, in the possession of

the Clockmaker's Company, which are still *actually kept going*. The number of watches manufactured for home consumption was, in the year 1798, about 50,000 annually. If this supply was for Great Britain only, it was consumed by about ten and a half millions of persons.

346. Machines are, in some trades, let out to hire, and a certain sum is paid for their use, in the manner of rent. This is the case amongst the framework knitters: and Mr Henson, in speaking of the rate of payment for the use of their frames, states, that the proprietor receives such a rent that, besides paying the full interest for his capital, he clears the value of his frame in nine years. When the rapidity with which improvements succeed each other is considered, this rent does not appear exorbitant. Some of these frames have been worked for thirteen years with little or no repair. But circumstances occasionally arise which throw them out of employment, either temporarily or permanently. Some years since, an article was introduced called *cut-up work*, by which the price of stocking-frames was greatly deteriorated. From the evidence of Mr J. Rawson, it appears that, in consequence of this change in the nature of the work, *each frame could do the work of two*, and many stocking frames were thrown out of employment, and their value reduced *full three-fourths*.* /

This information is of great importance, if the numbers here given are nearly correct, and if no other causes intervened to diminish the price of frames; for it shows the numerical connection between the increased production of those machines and their diminished value.

347. The great importance of simplifying all transactions between masters and workmen, and of dispassionately discussing with the latter the influence of any proposed regulations connected with their trade, is well examplified by a mistake into which both parties unintentionally fell, and which was productive of very great misery in the lace trade. Its history is so well told by William Allen, a framework knitter, who was a party to it, that an extract from his evidence, as given before the Framework Knitters' Committee of 1812, will best explain it.

I beg to say a few words respecting the frame rent; the rent paid for lace frames, until the year 1805, was 1*s*. 6*d*. a frame per week; there then was not any very great inducement for persons to buy frames and let them out by the

* Report from the Committee of the House of Commons on the Framework Knitter's Petition, April, 1819.

hire, who did not belong to the trade; at that time an attempt was made, by one or two houses, to reduce the prices paid to the workmen, in consequence of a dispute between these two houses and another great house: some little difference being paid in the price amongst the respective houses, I was one chosen by the workmen to try if we could not remedy the impending evil: we consulted the respective parties, and found them inflexible; these two houses that were about to reduce the prices, said that they would either immediately reduce the price of making net, or they would increase the frame rent: the difference to the workmen was considerable, between the one and the other; they would / suffer less, in the immediate operation of the thing, by having the rent advanced, than the price of making net reduced. They chose at that time, as they thought, the lesser evil, but it has turned out to be otherwise; for, immediately as the rent was raised upon the percentage laid out in frames, it induced almost every person, who had got a little money, to lay it out in the purchase of frames; these frames were placed in the hands of men who could get work for them at the warehouses; they were generally constrained to pay an enormous rent, and then they were compelled, most likely, to buy of the persons that let them the frames, their butcher's meat, their grocery, or their clothing: the encumbrance of these frames became entailed upon them: if any deadness took place in the work they must take it at a very reduced price, for fear of the consequences that would fall upon them from the person who bought the frame: thus the evil has been daily increasing, till, in conjunction with the other evils crept into the trade, they have almost crushed it to atoms.

348. The evil of not assigning fairly to each tool, or each article produced, its *proportionate value*, or even of not having a perfectly distinct, simple, and definite *agreement* between a master and his workmen, is very considerable. Workmen find it difficult in such cases to know the probable produce of their labour; and both parties are often led to adopt arrangements, which, had they been well examined, would have been rejected as equally at variance in the results with the true interests of both.

349. At Birmingham, stamps and dies, and presses for a great variety of articles, are let out: they are generally made by men possessing small capital, and are rented by workmen. Power also is rented at the same place. Steam-engines are / erected in large buildings containing a variety of rooms, in which each person may hire one, two, or any other amount of horsepower, as his occupation may require. If any mode could be discovered of transmitting power, without much loss from friction, to considerable distances, and at the same time of registering the quantity made use of at any particular point, a

considerable change would probably take place in many departments of the present system of manufacturing. A few central engines to produce power, might then be erected in our great towns, and each workman, hiring a quantity of power sufficient for his purpose, might have it conveyed into his own house; and thus a transition might in some instances be effected, if it should be found more profitable, back again from the system of great factories to that of domestic manufacture.

350. The transmission of water through a series of pipes, might be employed for the distribution of power, but the friction would consume a considerable portion. Another method has been employed in some instances, and is practised at the Mint. It consists in exhausting the air from a large vessel by means of a steam-engine. This vessel is connected by pipes, with a small piston which drives each coining press; and, on opening a valve, the pressure of the external air forces in the piston. This air is then admitted to the general reservoir, and pumped out by the engine. The condensation of air might be employed for the same purpose; but there are some unexplained facts relating to elastic fluids, which require further observations and experiment before they can be used for the conveyance / of power to any considerable distance. It has been found, for instance, in attempting to blow a furnace by means of a powerful water-wheel driving air through a cast-iron pipe of above a mile in length, that scarcely any sensible effect was produced at the opposite extremity. In one instance, some accidental obstruction being suspected, a cat put in at one end found its way out without injury at the other, thus proving that the phenomenon did not depend on interruption within the pipe.

351. The most portable form in which power can be condensed is, perhaps, by the liquefaction of the gases. It is known that, under considerable pressure, several of these become liquid at ordinary temperatures; carbonic acid, for example, is reduced to a liquid state by a pressure of sixty atmospheres. One of the advantages attending the use of these fluids, would be that the pressure exerted by them would remain constant until the last drop of liquid had assumed the form of gas. If either of the elements of common air should be found to be capable of reduction to a liquid state before it unites into a corrosive fluid with the other ingredient, then we shall possess a ready means of conveying power in any quantity and to any distance. Hydrogen probably will require the strongest compressing force to

render it liquid, and may, therefore, possibly be applied where still greater condensation of power is wanted. In all these cases the condensed gases may be looked upon as springs of enormous force, which have been wound up by the exertion of power, and which will deliver the whole of it back again when required. These springs of nature differ in / some respects from the steel springs formed by our art; for in the compression of the natural springs a vast quantity of latent heat is forced out, and in their return to the state of gas an equal quantity is absorbed. May not this very property be employed with advantage in their application?

Part of the mechanical difficulty to be overcome in constructing apparatus connected with liquefied gases, will consist in the structure of the valves and packing necessary to retain the fluids under the great pressure to which they must be submitted. The effect of heat on these gases has not yet been sufficiently tried, to lead us to any very precise notions of the additional power which its application to them will supply.

The elasticity of air is sometimes employed as a spring, instead of steel: in one of the large printing-machines in London the momentum of a considerable mass of matter is destroyed by making it condense the air included in a cylinder, by means of a piston against which it impinges.

352. The effect of competition in cheapening articles of manufacture sometimes operates in rendering them less durable. When such articles are conveyed to a distance for consumption, if they are broken, it often happens, from the price of labour being higher where they are used than where they were made, that it is more expensive to mend the old article, than to purchase a new. Such is usually the case, in great cities, with some of the commoner locks, with hinges, and with a variety of articles of hardware. /

[293/4]

CHAPTER XXX

ON COMBINATIONS AMONGST MASTERS OR WORKMEN AGAINST EACH OTHER

353. There exist amongst the workmen of almost all classes, certain rules or laws which govern their actions towards each other, and towards their employers. But, besides these general principles, there are frequently others peculiar to each factory, which have derived their origin, in many instances, from the mutual convenience of the parties engaged in them. Such rules are little known except to those actually pursuing the several trades; and, as it is of importance that their advantages and disadvantages should be canvassed, we shall offer a few remarks upon some of them.

354. The principles by which such laws should be tried are,

First. *That they conduce to the general benefit of all the persons employed.*

Secondly. *That they prevent fraud.*

Thirdly. *That they interfere as little as possible with the free agency of each individual.*

355. It is usual in many workshops, that, on the first entrance of a new journeyman, he shall pay a small fine to the rest of the men. It is clearly unjust to insist upon this payment; and when it is spent / in drinking, which is, unfortunately, too often the case, it is injurious. The reason assigned for the demand is, that the newcomer will require some instruction in the habits of the shop, and in the places of the different tools, and will thus waste the time of some of his companions until he is instructed. If this fine were added to a fund, managed by the workmen themselves, and either divided at given periods, or reserved for their relief in sickness, it would be less objectionable, since its tendency would be to check the too frequent change of men from one shop to another. But it ought, at all events, not to be compulsory; and the advantages to be derived from the fund to which the workman is invited to subscribe, ought to be his sole inducement to contribute.

356. In many workshops, the workmen, although employed on totally different parts of the objects manufactured, are yet dependent, in some measure, upon each other. Thus a single smith may be able to forge, in one day, work enough to keep four or five turners employed during the next. If, from idleness or intemperance, the smith neglects his work, and does not furnish the usual supply, the turners (supposing them to be paid by the piece), will have their time partly unoccupied, and their gains consequently diminished. It is reasonable, in such circumstances, that a fine should be levied on the delinquent; but it is desirable that the master should have concurred with his workmen in establishing such a rule, and that it should be shown to each individual previously to his engagement; and it is very desirable that such fine should not be spent in drinking. /

357. In some establishments, it is customary for the master to give a small gratuity whenever any workman has exercised a remarkable degree of skill, or has economized the material employed. Thus, in splitting horn into layers for lanterns, one horn usually furnishes from five to eight layers; but if a workman split the horn into ten layers or more, he receives a pint of ale from the master. These premiums should not be too high, lest the material should be wasted in unsuccessful attempts: but such regulations, when judiciously made, are beneficial, as they tend to produce skill amongst the workmen, profit to the masters, and diminished cost to the consumers.

358. In some few factories, in which the men are paid by the piece, it is usual, when any portion of work, delivered in by a workman, is rejected by the master on account of its being badly executed, to fine the delinquent. Such a practice tends to remedy one of the evils attendant upon that mode of payment, and greatly assists the master, since his own judgement is thus supported by competent and unprejudiced judges.

359. Societies exist amongst some of the larger bodies of workmen, and others have been formed by the masters engaged in the same branches of trade. These associations have different objects in view; but it is very desirable that their effects should be well understood by the individuals who compose them; and that the advantages arising from them, which are certainly great, should be separated as much as possible from the evils which they have, unfortunately, too frequently introduced. Associations of workmen and of masters may, with advantage, agree / upon rules to be observed by both parties, in estimating the proportionate value of different kinds of work

executed in their trade, in order that time may be saved, and disputes be prevented. They may also be most usefully employed in acquiring accurate information as to the number of persons working in the various departments of any manufacture, their rate of wages, the number of machines in use, and other statistical details. Information of this nature is highly valuable, both for the guidance of the parties who are themselves most interested, and to enable them, upon any application to government for assistance, or with a view to legislative enactments, to supply those details, without which the propriety of any proposed measure cannot be duly estimated. Such details may be collected by men actually engaged in any branch of trade, at a much smaller expense of time, than by persons less acquainted with, and less interested in it.

360. One of the most legitimate and most important objects of such associations as we have just mentioned, is to agree upon ready and certain modes of measuring the quantity of work done by the workmen. For a long time a difficulty upon this point existed in the lace trade, which was justly complained of by the men as a serious grievance; but the introduction of the *rack*, which counts the number of holes in the length of the piece, has entirely put an end to the most fertile cause of disputes. This invention was adverted to by the Committee of 1812, and a hope was expressed, in their report, that the same contrivance would be applied to / stocking-frames. It would, indeed, be of great mutual advantage to the industrious workman, and to the master manufacturer in every trade, if the machines employed in it could register the quantity of work which they perform, in the same manner as a steam-engine does the number of strokes it makes. The introduction of such contrivances gives a greater stimulus to honest industry than can readily be imagined, and removes one of the sources of disagreement between parties, whose real interests must always suffer by any estrangement between them.

361. The effects arising from combinations amongst the workmen, are almost always injurious to the parties themselves. There are numerous instances, in which the public suffer by increased price at the moment, but are ultimately gainers from the permanent reduction which results; whilst, on the other hand, the improvements which are often made in machinery in consequence of 'a strike' amongst the workmen, most frequently do injury, of greater or less duration, to that particular class which gave rise to them. As the injury to the men and to their families is almost always more serious

than that which affects their employers, it is of the utmost importance to the comfort and happiness of the former class, that they should themselves entertain sound views upon this question. For this purpose a few illustrations of the principle which is here maintained, will probably have greater weight than any reasoning of a more general nature, though drawn from admitted principles of political economy. Such instances will, moreover, present the advantage of / appealing to facts known to many individuals of those classes for whose benefit these reflections are intended.

362. There is a process in the manufacture of gun barrels for making what, in the language of the trade, are called *skelps*. The *skelp* is a piece or bar of iron, about three feet long, and four inches wide, but thicker and broader at one end than at the other: and the barrel of a musket is formed by forging out such pieces to the proper dimensions, and then folding or bending them into a cylindrical form, until the edges overlap, so that they can be welded together.

About twenty years ago, the workmen, employed at a very extensive factory in forging these skelps out of bar-iron, 'struck' for an advance of wages; and as their demands were very exorbitant, they were not immediately complied with. In the meantime, the superintendent of the establishment directed his attention to the subject; and it occurred to him, that if the circumference of the rollers, between which the bar-iron was rolled, were to be made equal to the length of a *skelp*, or of a musket barrel, and if also the groove in which the iron was compressed, instead of being of the same width and depth throughout, were cut gradually deeper and wider from a point on the rollers, until it returned to the same point, then the bar-iron passing between such rollers, instead of being uniform in width and thickness, would have the form of a *skelp*. On making the trial, it was found to succeed perfectly; a great reduction of human labour was effected by the process, and the workmen who had acquired / peculiar skill in performing it ceased to derive any advantage from their dexterity.

363. It is somewhat singular that another and a still more remarkable instance of the effect of combination amongst workmen, should have occurred but a few years since in the very same trade. The process of welding the *skelps*, so as to convert them into gun barrels, required much skill, and after the termination of the war, the demand for muskets having greatly diminished, the number of persons employed in making them was very much reduced. This circumstance rendered

combination more easy; and upon one occasion, when a contract had been entered into for a considerable supply to be delivered on a fixed day, the men all struck for such an advance of wages as would have caused the completion of the contract to be attended with a very heavy loss.

In this difficulty, the contractors resorted to a mode of welding the gun barrel, for which a patent had been taken out by one of themselves some years before this event. The plan had not then succeeded so well as to come into general use, in consequence of the cheapness of the usual mode of welding by hand labour, combined with some other difficulties with which the patentee had to contend. But the stimulus produced by the combination of the workmen, induced him to make new trials, and he was enabled to introduce such a facility in welding gun barrels by rollers, and such perfection in the work itself, that, in all probability, very few will in future be welded by hand labour.

This new process consisted in folding a bar of iron, about a foot long, into the form of a cylinder, with / the edges a little overlapping. It was then placed in a furnace, and being taken out when raised to a welding heat, a triblet, or cylinder of iron, was placed in it, and the whole was passed quickly through a pair of rollers. The effect of this was, that the welding was performed at a single heating, and the remainder of the elongation necessary for extending the *skelps* to the length of the musket barrel, was performed in a similar manner, but at a lower temperature. The workmen who had combined were, of course, no longer wanted, and instead of benefiting themselves by their combination, they were reduced permanently, by this improvement in the art, to a considerably lower rate of wages: for as the process of welding gun barrels by hand required peculiar skill and considerable experience, they had hitherto been in the habit of earning much higher wages than other workmen of their class. On the other hand, the new method of welding was far less injurious to the texture of the iron, which was now exposed only once, instead of three or four times, to the welding heat, so that the public derived advantage from the superiority, as well as from the economy of the process. Another process has subsequently been invented, applicable to the manufacture of a lighter kind of iron tubes, which can thus be made at a price which renders their employment very general. They are now to be found in the shops of all our larger ironmongers, of various lengths and diameters, with screws cut at each end; and are in constant use for the conveyance of gas for lighting, or of water for warming, our houses. /

364. Similar examples must have presented themselves to all those who are familiar with the details of our manufactories, but these are sufficient to illustrate one of the results of combinations. It would not, however, be fair to push the conclusion deduced from these instances to its extreme limit. Although it is very apparent, that in the two cases which have been stated, the effects of combination were permanently injurious to the workman, by almost immediately placing him in a lower class (with respect to his wages) than he occupied before; yet they do not prove that *all* such combinations have this *effect*. It is quite evident that they have all this *tendency*; it is also certain that considerable stimulus must be applied to induce a man to contrive a new and expensive process; and that in both these cases, unless the fear of pecuniary loss had acted powerfully, the improvement would not have been made. If, therefore, the workmen had in either case combined for only a small advance of wages, they would, in all probability, have been successful, and the public would have been deprived, for many years, of the inventions to which these combinations gave rise. It must, however, be observed, that the same skill which enabled the men to obtain, after long practice, higher wages than the rest of their class, would prevent many of them from being *permanently* thrown back into the class of ordinary workmen. Their diminished wages will continue only until they have acquired, by practice, a facility of execution in some other of the more difficult operations: but a diminution of wages, even for a year or two, is still a very serious inconvenience to / any person who lives by his daily exertion. The consequence of combination has then, in these instances, been, to the workmen who combined – reduction of wages; to the public – reduction of price; and to the manufacturer – increased sale of his commodity, resulting from that reduction.

365. It is, however, important to consider the effects of combination in another and less obvious point of view. The fear of combination amongst the men whom he employs, will have a tendency to induce the manufacturer to conceal from his workmen the extent of the orders he may at any time have received; and, consequently, they will always be less acquainted with the extent of the demand for their labour than they otherwise might be. This is injurious to their interests; for instead of foreseeing, by the gradual falling-off in the orders, the approach of a time when they must be unemployed, and

preparing accordingly, they are liable to much more sudden changes than those to which they would otherwise be exposed.

In the evidence given by Mr Galloway, the engineer, he remarks, that,

> When employers are competent to show their men that their business is steady and certain, and when men find that they are likely to have permanent employment, they have always better habits, and more settled notions, which will make them better men, and better workmen, and will produce great benefits to all who are interested in their employment.

366. As the manufacturer, when he makes a contract, has no security that a combination may not arise amongst the workmen, which may render that contract a loss instead of a benefit; besides / taking precautions to prevent them from becoming acquainted with it, he must also add to the price at which he could otherwise sell the article, some small increase to cover the risk of such an occurrence. If an establishment consist of several branches which can only be carried on jointly, as, for instance, of iron mines, blast furnaces, and a colliery, in which there are distinct classes of workmen, it becomes necessary to keep on hand a larger stock of materials than would be required, if it were certain that no combinations would arise.

Suppose, for instance, the colliers were to 'strike' for an advance of wages – unless there was a stock of coal above ground, the furnaces must be stopped, and the miners also would be thrown out of employ. Now the cost of keeping a stock of iron ore, or of coals above ground, is just the same as that of keeping in a drawer, unemployed, its value in money, (except, indeed, that the coal suffers a small deterioration by exposure to the elements). The interest of this sum must, therefore, be considered as the price of an insurance against the risk of combination amongst the workmen; and it must, so far as it goes, increase the price of the manufactured article, and, consequently, limit the demand which would otherwise exist for it. But every circumstance which tends to limit the demand, is injurious to the workmen; because the wider the demand, the less it is exposed to fluctuation.

The effect to which we have alluded, is by no means a theoretical conclusion; the proprietors of one establishment in the iron trade, within the author's knowledge, think it expedient always to keep above ground a supply of coal for six months, which / is, in that instance, equal in value to about £10,000. When we reflect that the quantity of capital throughout the country thus kept unemployed merely from the fear of *combinations amongst the workmen*, might, under other circumstances, be used for keeping a larger number at work, the importance of introducing a system in which there should exist no inducement to combine becomes additionally evident.

367. That combinations are, while they last, productive of serious inconveniences to the workmen themselves, is admitted by all parties; and it is equally true, that, in most cases, a successful result does not leave them in so good a condition as they were in before 'the strike'. The little capital they possessed, which ought to have been hoarded with care for days of illness or distress, is exhausted; and frequently, in order to gratify a pride, at the existence of which we cannot but rejoice, even whilst we regret its misdirected energy, they will undergo the severest privations rather than return to work at their former wages. With many of the workmen, unfortunately, during such periods, bad habits are formed which it is very difficult to eradicate; and, in all those engaged in such transactions, the kinder feelings of the heart are chilled, and passions are called into action which are permanently injurious to the happiness of the individual, and destructive of those sentiments of confidence which it is equally the interest of the master manufacturer and of his workman to maintain. If any of the trade refuse to join in the strike, the majority too frequently forget, in the excitement of their feelings, the dictates of justice, and endeavour to exert a / species of tyranny, which can never be permitted to exist in a free country. In conceding therefore to the working classes, that they have a *right*, if they consider it expedient, to combine for the purpose of procuring higher wages (provided always, that they have completed all their existing contracts), it ought ever to be kept before their attention, that the same freedom which they claim for themselves they are bound to allow to others, who may have different views of the advantages of combination. Every effort which reason and kindness can dictate, should be made, not merely to remove their grievances, but to satisfy their own reason and feelings, and to show them the consequences which will probably result from their conduct: but the strong arm of the law, backed, as in such cases it will always be, by public opinion, should be instantly and unhesitatingly applied, to prevent them from violating the liberty of a portion of their own, or of any other class of society.

368. Amongst the evils which ultimately fall heavy on the working

classes themselves, when, through mistaken views, they attempt to interfere with their employers in the mode of carrying on their business, may be mentioned the removal of factories to other situations, where the proprietors may be free from the improper control of their men. The removal of a considerable number of lace frames to the western counties, which took place, in consequence of the combinations in Nottinghamshire, has already been mentioned. Other instances have occurred, where still greater injury has been produced by the removal of a portion of the skill and capital of the country to a foreign land. Such was the case at / Glasgow, as stated in the fifth Parliamentary Report respecting Artizans and Machinery. One of the partners in an extensive cotton factory, disgusted by the unprincipled conduct of the workmen, removed to the state of New York, where he re-established his machinery, and thus afforded, to rivals already formidable to our trade, at once a pattern of our best machinery, and an example of the most economical methods of employing it.

369. When the nature of the work is such that it is not possible to remove it, as happens with regard to mines, the proprietors are more exposed to injury from combinations amongst the workmen: but as the owners are generally possessed of a larger capital, they generally succeed, if the reduction of wages which they propose is really founded on the necessity of the case.

An extensive combination lately existed amongst the colliers in the north of England, which unfortunately led, in several instances, to acts of violence. The proprietors of the coalmines were consequently obliged to procure the aid of miners from other parts of England who were willing to work at the wages they could afford to give; and the aid of the civil, and in some cases of the military, power, was requisite for their protection. This course was persisted in during several months, and the question being, which party could support itself longest on the diminished gains, as it might have readily been foreseen, the proprietors ultimately succeeded.

370. One of the remedies employed by the masters against the occurrence of combinations, is to make engagements with their men for long periods and to / arrange them in such a manner, that these contracts shall not all terminate together. This has been done in some cases at Sheffield, and in other places. It is attended with the inconvenience to the masters that, during periods when the demand for their produce is reduced, they are still obliged to employ the same

number of workmen. This circumstance, however, frequently obliges the proprietors to direct their attention to improvements in their works; and in one such instance, within the author's knowledge, a large reservoir was deepened, thus affording a more constant supply to the water-wheel, whilst, at the same time, the mud from the bottom gave permanent fertility to a piece of land previously almost barren. In this case, not merely was the supply of produce checked, when a glut existed, but the labour was, in fact, applied more profitably than it would have been in the usual course.

371. A mode of paying the wages of workmen in articles which they consume, has been introduced into some of our manufacturing districts, which has been called the *truck system.* As in many instances this has nearly the effect of a combination of the masters against the men, it is a fit subject for discussion in the present chapter: but it should be carefully distinguished from another system of a very different tendency, which will be first described.

372. The principal necessaries for the support of a workman and his family are few in number, and are usually purchased by him in small quantities weekly. Upon such quantities, sold by the retail dealer, a large profit is generally made; and if the article is one whose quality, like that of tea, is not readily / estimated, then a great additional gain is made by the retail dealer selling an inferior article.

Where the number of workmen living on the same spot is large, it may be thought desirable that they should unite together and have an agent, to purchase by wholesale those articles which are most in demand, as tea, suger, bacon, etc., and to retail them at prices, which will just repay the wholesale cost, together with the expense of the agent who conducts their sale. If this be managed wholly by a committee of workmen, aided perhaps by advice from the master, and if the agent is paid in such a manner as to have himself an interest in procuring good and reasonable articles, it may be a benefit to the workmen: and if the plan succeed in reducing the cost of articles of necessity to the men, it is clearly the interest of the master to encourage it. The master may indeed be enabled to afford them facilities in making their wholesale purchases; but he ought never to have the least interest in, or any connection with, the profit made by the articles sold. The men, on the other hand, who subscribe to set up the shop, ought not, in the slightest degree, to be compelled to make their purchases there: the goodness and cheapness of the article ought to be their sole inducements.

It may perhaps be objected, that this plan is only employing a portion of the capital belonging to the workmen in a retail trade; and that, without it, competition amongst small shopkeepers will reduce the articles to nearly the same price. This objection would be valid if the objects of consumption required no *verification*; but combining what has been already / stated on that subject* with the present argument, the plan seems liable to no serious objections.

373. The *truck system* is entirely different in its effects. The master manufacturer keeps a retail shop for articles required by his men, and either pays their wages in goods, or compels them by express agreement, or less directly, by unfair means, to expend the whole or a certain part of their wages at his shop. If the manufacturer kept this shop merely for the purpose of securing good articles, at fair prices, to his workmen, and if he offered no inducement to them to purchase at his shop, except the superior cheapness of his articles, it would certainly be advantageous to the men. But, unfortunately, this is not always the case; and the temptation to the master, in times of depression, to reduce in effect the wages which he pays (by increasing the price of articles at his shop), without altering the nominal rate of payment, is frequently too great to be withstood. If the object be solely to procure for his workmen better articles, it will be more effectually accomplished by the master confining himself to supplying a small capital, at a moderate rate of interest; leaving the details to be conducted by a committee of workmen, in conjunction with his own agent, and the books of the shop to be audited periodically by the men themselves.

374. Wherever the workmen are paid in goods, or are compelled to purchase at the master's shop, much injustice is done to them, and great misery results from it. Whatever may have been the intentions / of the master in such cases, the real effect is, to deceive the workman as to the amount he receives in exchange for his labour. Now, the principles on which the happiness of that class of society depends, are difficult enough to be understood, even by those who are blessed with far better opportunities of investigating them: and the importance of their being well acquainted with those principles which relate to themselves, is of more vital consequence to workmen, than to many other classes. It is therefore highly desirable to assist them in comprehending the position in which they are placed, by rendering

* See Chapter XV, p. 87.

all the relations in which they stand to each other, and to their employers, as simple as possible. Workmen should be paid entirely in money; their work should be measured by some unbiassed, some unerring piece of mechanism; the time during which they are employed should be defined, and punctually adhered to. The payments they make to their benefit societies should be fixed on such just principles, as not to require extraordinary contributions. In short, the object of all who wish to promote their happiness should be, to give them, in the simplest form, the means of knowing beforehand, the sum they are likely to acquire by their labour, and the money they will be obliged to expend for their support: thus putting before them, in the clearest light, the certain result of persevering industry.

375. The cruelty which is inflicted on the workman by the payment of his wages in goods, is often very severe. The little purchases necessary for the comfort of his wife and children, perhaps the medicines he occasionally requires for them in illness, / must all be made through the medium of barter; and he is obliged to waste his time in arranging an exchange, in which the goods which he has been compelled to accept for his labour are invariably taken at a lower price than that at which his master charged them to him. The father of a family perhaps, writhing under the agonies of the toothache, is obliged to make his hasty bargain with the village surgeon, before he will remove the cause of his pain; or the disconsolate mother is compelled to sacrifice her depreciated goods in exchange for the last receptacle of her departed offspring. The subjoined evidence from the Report of the Committee of the House of Commons on Framework Knitters' Petitions, shows that these are not exaggerated statements.

It has been so common in our town to pay goods instead of money, that a number of my neighbours have been obliged to pay articles for articles, to pay sugar for drugs out of the druggist's shop; and others have been obliged to pay sugar for drapery goods, and such things, and exchange in that way numbers of times. I was credibly informed, that one person paid half a pound of tenpenny sugar and a penny to have a tooth drawn; and there is a credible neighbour of mine told me, that he had heard that the sexton had been paid for digging a grave with sugar and tea; and before I came off, knowing I had to give evidence upon these things, I asked this friend to enquire of the sexton, whether this was a fact: the sexton hesitated for a little time, on account of bringing into discredit the person who paid these goods; however, he said at last, 'I have received these articles repeatedly – I know these things have been paid to a great extent in this way.' /

[312/3]

CHAPTER XXXI

ON COMBINATIONS OF MASTERS AGAINST THE PUBLIC

376. A species of combination occasionally takes place amongst manufacturers against persons having patents: and these combinations are always injurious to the public, as well as unjust to the inventors. Some years since, a gentleman invented a machine, by which modellings and carvings were cut in mahogany, and other fine woods. The machine resembled, in some measure, the drilling apparatus employed in ornamental lathes; it produced beautiful work at a very moderate expense: but the cabinetmakers met together, and combined against it, and the patent has consequently never been *worked.* A similar fate awaited a machine for cutting veneers by means of a species of knife. In this instance, the wood could be cut thinner than by the circular saw, and no waste was incurred; but 'the trade' set themselves against it, and after a heavy expense, it was given up.

The excuse alleged for this kind of combination, was the fear entertained by the cabinetmakers that when the public became acquainted with the article, the patentee would raise the price.

Similar examples of combination seem not to be unfrequent, as appears by the Report of the Committee of the House of Commons on Patents for Inventions, June, 1829. See the evidence of Mr Holdsworth.

377. There occurs another kind of combination against the public, with which it is difficult to deal. It / usually ends in a monopoly, and the public are then left to the discretion of the monopolists not to charge them above the *growling point* – that is, *not to make them pay so much as to induce them actually to combine against the imposition.* This occurs when two companies supply water or gas to consumers by means of pipes laid down under the pavement in the street of cities: it may possibly occur also in docks, canals, railroads, etc., and in other cases where the capital required is very large, and the competition very limited. If water or gas companies combine, the public immediately

loses all the advantage of competition, and it has generally happened, that at the end of a period during which they have undersold each other, the several companies have agreed to divide the whole district supplied, into two or more parts, each company then removing its pipes from all the streets except those in its own portion. This removal causes great injury to the pavement, and when the pressure of increased rates induces a new company to start, the same inconvenience is again produced. Perhaps one remedy against evils of this kind might be, when a charter is granted to such companies, to restrict, to a certain amount, the rate of profit on the shares, and to direct that any profits beyond, shall accumulate for the repayment of the original capital. This has been done in several late Acts of Parliament establishing companies. The maximum rate of profit allowed ought to be liberal, to compensate for the risk; the public ought to have auditors on their part, and the accounts should be annually published, for the purpose of preventing the limitations from being exceeded. It must however / be admitted, that this would be an interference with capital, which, if allowed, should, in the present state of our knowledge, be examined with great circumspection in each individual case, until some general principle is established on well-admitted grounds.

378. An instrument called a gas-meter, which ascertains the quantity of gas used by each consumer, has been introduced, and furnishes a satisfactory mode of determining the payments to be made by individuals to the gas companies. A contrivance somewhat similar in its nature, might be used for the sale of water; but in that case some public inconvenience might be apprehended, from the diminished quantity which would then run to waste: the streams of water running through the sewers in London, are largely supplied from this source; and if this supply were diminished, the drainage of the metropolis might be injuriously affected.

379. In the north of England a powerful combination has long existed among the coal-owners, by which the public has suffered in the payment of increased price. The late examination of evidence before a Committee of the House of Commons, has explained its mode of operation, and the Committee have recommended, that for the present the sale of coal should be left to the competition of other districts.

380. A combination, of another kind, exists at this moment to a great extent, and operates upon the price of the very pages which are now communicating information respecting it. A subject so interesting to

every reader, and still more so to every manufacturer of the article which the reader consumes, deserves an attentive examination. /

We have shown in Chapter XXI, p. 144, the component parts of the expense of each copy of the present work; and we have seen that the total amount of the cost of its production, exclusive of any payment to the author for his labour, is 2*s*. 3*d*.*

Another fact, with which the reader is more practically familiar, is that he has paid, or is to pay, to his bookseller, six shillings for the volume. Let us now examine into the distribution of these six shillings, and then, having the facts of the case before us, we shall be better able to judge of the merits of the combination just mentioned, and to explain its effects.

Distribution of the profits on a six shilling book

	Buys at	*Sells at*	*Profit on capital expended*
	s. d.	*s. d.*	
No. I – *The publisher* who accounts to the author for every copy received	3 10	4 2	10 per cent
No. II – *The bookseller* who retails to the public	4 2	6 0	44
Or,	4 6	6 0	33⅓

No. I, the *publisher*, is a bookseller; he is, in fact, the author's agent. His duties are, to receive and take charge of the stock, for which he supplies warehouse room; to advise the author about the times and methods of advertising; and to insert the advertisements. As he publishes other books, he / will advertise lists of those sold by himself; and thus, by combining many in one advertisement, diminish the expense to each of his principals. He pays the author only for the books actually sold; consequently, he makes no outlay of capital, except that which he pays for advertisements: but he is answerable for any bad debts he may contract in disposing of them. His charge is usually ten per cent on the returns.

No. II is the *bookseller* who retails the work to the public. On the publication of a new book, the *publisher* sends round to the trade, to receive 'subscriptions' from them for any number of copies not less than two. These copies are usually charged to the 'subscribers', on an average, at about four or five per cent less than the wholesale price of the

* The whole of the subsequent details relate to the *first* edition of this work.

book: in the present case the subscription price is 4*s*. 2*d*. for each copy. After the day of publication, the price charged by the publisher to the booksellers is 4*s*. 6*d*. With some works it is the custom to deliver twenty-five copies to those who order twenty-four, thus allowing a reduction of about four per cent. Such was the case with the present volume. Different publishers offer different terms to the subscribers; and it is usual, after intervals of about six months, for the publisher again to open a subscription list, so that if the work be one for which there is a steady sale, the trade avail themselves of these opportunities of purchasing, at the reduced rate, enough to supply their probable demand.*

381. The volume thus purchased of the publisher / at 4*s*. 2*d*. or 4*s*. 6*d*. is retailed by the bookseller to the public at 6*s*. In the first case he makes a profit of forty-four, in the second of thirty-three per cent. Even the smaller of these two rates of profit on the capital employed, appears to be much too large. It may sometimes happen, that when a book is enquired for, the retail dealer sends across the street to the wholesale agent, and receives, for this trifling service, one fourth part of the money paid by the purchaser; and perhaps the retail dealer takes also six months' credit for the price which the volume actually cost him.

382. In section 256, the price of each process in manufacturing the present volume was stated: we shall now give an analysis of the whole expense of conveying it into the hands of the public.

	£	s.	d.	£	s.	d.
The retail price 6*s*. on 3052 produces				915	12	0
1 Total expense of printing and paper				207	5	8⁷⁄11
2 Taxes on paper and advertisements				40	0	11
3 Commission to publisher as agent between author and printer	18	14	4⁴⁄11			
4 Commission to publisher as agent for sale of the book	63	11	8			
5 Profit – the difference between subscription price and trade price, 4*d*. per vol.	50	17	4			
6 Profit – the difference between trade price and retail price, 1*s*. 6*d*. per vol.	228	18	0			
				362	1	4 /11
7 Remains for authorship				306	4	0
Total				915	12	0 /

* These details vary with different books and different publishers; those given in the text are believed to be substantially correct, and are applicable to works like the present.

This account appears to disagree with that in page [206] 146; but it will be observed that the three first articles amount to £266 1*s*., the sum there stated. The apparent difference arises from a circumstance which was not noticed in the first edition of this work. The bill amounting to £205 18*s*., as there given, and as reprinted in the present volume, included an additional charge of ten per cent upon the real charges of the printer and paper-maker.

383. It is usual for the publisher, when he is employed as agent between the author and printer, to charge a commission of ten per cent on all payments he makes. If the author is informed of this custom previously to his commencing the work, as was the case in the present instance, he can have no just cause of complaint; for it is optional whether he himself employs the printer, or communicates with him through the intervention of his publisher.

The services rendered for this payment are, the making arrangements with the printer, the wood-cutter, and the engraver, if required. There is a convenience in having some intermediate person between the author and printer, in case the former should consider any of the charges made by the latter as too high. When the author himself is quite unacquainted with the details of the art of printing, he may object to charges which, on a better acquaintance with the subject, he might be convinced were very moderate; and in such cases he ought to depend on the judgement of his publisher, who is generally conversant with the art. This is particularly the case in the charge for alterations and corrections, some of which, although apparently trivial, occupy / the compositors much time in making. It should also be observed that the publisher, in this case, becomes responsible for the payments to those persons.

384. It is not necessary that the author should avail himself of this intervention, although it is the interest of the publisher that he should; and booksellers usually maintain that the author cannot procure his paper or printing at a cheaper rate if he go at once to the producers. This appears from the evidence given before the Committee of the House of Commons in the Copyright Acts, 8 May, 1818.

Mr O. Rees, bookseller, of the house of Longman and Co., Paternoster Row, examined:

Q. Suppose a gentleman to publish a work on his own account, and to incur all the various expenses; could he get the paper at 30*s*. a ream?

A. I presume not; I presume a stationer would not sell the paper at the same price to an indifferent gentleman as to the trade.

Q. The Committee asked you if a private gentleman was to publish a work on his own account, if he would not pay more for the paper than persons in the trade; the Committee wish to be informed whether a printer does not charge a gentleman a higher rate than to a publisher.

A. I conceive they generally charge a profit on the paper.

Q. Do not the printers charge a higher price also for printing, than they do to the trade?

A. I always understood that they do.

385. There appears to be little reason for this distinction in charging for printing a larger price to the author than to the publisher, provided the former is able to give equal security for the payment. With respect to the additional charge on paper, if the / author employs either publisher or printer to purchase it, they ought to receive a moderate remuneration for the risk, since they become responsible for the payment; but there is no reason why, if the author deals at once with the paper-maker, he should not purchase on the same terms as the printer; and if he choose, by paying ready money, not to avail himself of the long credit allowed in those trades, he ought to procure his paper considerably cheaper.

386. It is time, however, that such conventional combinations between different trades should be done away with. In a country so eminently depending for its wealth on its manufacturing industry, it is of importance that there should exist no *abrupt* distinction of classes, and that the highest of the aristocracy should feel proud of being connected, either personally or through their relatives, with those pursuits on which their country's greatness depends. The wealthier manufacturers and merchants already mix with those classes, and the larger and even the middling tradesmen are frequently found associating with the gentry of the land. It is good that this ambition should be cultivated, not by any rivalry in expense, but by a rivalry in knowledge and in liberal feelings; and few things would more contribute to so desirable an effect, than the abolition of all such contracted views as those to which we have alluded. The advantage to the other classes, would be an increased acquaintance with the productive arts of the country – an increased attention to the importance of acquiring habits of punctuality and of business – and, above all, a general feeling that it is honourable, in any rank of life, to increase our own and our country's riches, by / employing our talents in the production or in the distribution of wealth.

387. Another circumstance omitted to be noticed in the first edition

relates to what is technically called *the overplus*, which may be now explained. When 500 copies of a work are to be printed, each sheet of it requires one ream of paper. Now a ream, as used by printers, consists of 21½ quires, or 516 sheets. This excess of sixteen sheets is necessary in order to allow for 'revises' – for preparing and adjusting the press for the due performance of its work, and to supply the place of any sheets which may be accidentally dirtied or destroyed in the processes of printing, or injured by the binder in putting into boards. It is found, however, that three per cent is more than the proportion destroyed, and that damage is less frequent in proportion to the skill and care of the workmen.

From the evidence of several highly respectable booksellers and printers, before the Committee of the House of Commons on the Copyright Act, May, 1818, it appears that the average number of surplus copies, above 500, is between two and three; that on smaller impressions it is less, whilst on larger editions it is greater; that, in some instances, the complete number of 500 is not made up, in which case the printer is obliged to pay for completing it; and that in no instance have the whole sixteen extra copies been completed. On the volume in the reader's hands, the edition of which consisted of 3000, the surplus amounted to fifty-two – a circumstance arising from the improvements in printing and the increased care of the pressmen. Now this overplus ought to be / accounted for to the author – and I believe it usually is so by all respectable publishers.

388. In order to prevent the printer from privately taking off a larger number of impressions than he delivers to the author or publisher, various expedients have been adopted. In some works a particular watermark has been used in paper made purposely for the book: thus the words 'Mecanique Celeste' appear in the watermark of the two first volumes of the great work of Laplace. In other cases, where the work is illustrated by engravings, such a fraud would be useless without the concurrence of the copperplate printer. In France it is usual to print a notice on the back of the title page, that no copies are genuine without the subjoined signature of the author: and attached to this notice is the author's name, either written, or printed by hand from a wooden block. But notwithstanding this precaution, I have recently purchased a volume, printed at Paris, in which the notice exists, but no signature is attached. In London there is not much danger of such frauds, because the printers are men of capital, to whom the profit on such a transaction would be trifling, and the risk

of the detection of a fact, which must of necessity be known to many of their workmen, would be so great as to render the attempt at it folly.

389. Perhaps the best advice to an author, if he publishes on his own account, and is a reasonable person, possessed of common sense, would be to go at once to a respectable printer and make his arrangements with him.

390. If the author do not wish to print his / work at his own risk, then he should make an agreement with a publisher for an edition of a limited number; but *he should by no means sell the copyright.* If the work contains woodcuts or engravings, it would be judicious to make it part of the contract that they shall become the author's property, with the view to their use in a subsequent edition of the works, if they should be required. An agreement is frequently made by which the publisher advances the money and incurs all the risk on condition of his sharing the profits with the author. The profits alluded to are, for the present work, the last item of section 382, or £306 4*s*.

391. Having now explained all the arrangements in printing the present volume, let us return to section 382, and examine the distribution of the £915 paid by the public. Of this sum £207 was the cost of the book, £40 was taxes, £362 was the charges of the bookseller in conveying it to the consumer, and £306 remained for authorship.

The largest portion, or £362 goes into the pockets of the booksellers; and as they do not advance capital, and incur very little risk, this certainly appears to be an unreasonable allowance. The most extravagant part of the charge is the thirty-three per cent which is allowed as profit on retailing the book.

It is stated, however, that all retail booksellers allow to their customers a discount of ten per cent upon orders above 20*s*., and that consequently the nominal profit of forty-four or thirty-three per cent is very much reduced. If this is the case, it may fairly be enquired, why the price of £2 for / example, is printed upon the back of a book, when every bookseller is ready to sell it at £1 16*s*., and why those who are unacquainted with that circumstance should be made to pay more than others who are better informed?

392. Several reasons have been alleged as justifying this high rate of profit.

First, it has been alleged that the purchasers of books take long credit. This, probably, is often the case, and admitting it, no reasonable person can object to a proportionate increase of price.

But it is no less clear, that persons who do pay ready money, should not be charged the same price as those who defer their payments to a remote period.

Secondly, it has been urged that large profits are necessary to pay for the great expenses of bookselling establishments; that rents are high and taxes heavy; and that it would be impossible for the great booksellers to compete with the smaller ones, unless the retail profits were great. In reply to this it may be observed that the booksellers are subject to no peculiar pressure which does not attach to all other retail trades. It may also be remarked that large establishments always have advantages over smaller ones, in the economy arising from the division of labour; and it is scarcely to be presumed that booksellers are the only class who, in large concerns, neglect to avail themselves of them.

Thirdly, it has been pretended that this high rate of profit is necessary to cover the risk of the bookseller's having some copies left on his shelves; but he is not obliged to buy of the publisher a single copy more than he has orders for: and if he do purchase more, / at the subscription price, he proves, by the very fact, that he himself does not estimate that risk at more than from four to eight per cent.

393. It has been truly observed, on the other hand, that many copies of books are spoiled by persons who enter the shops of booksellers without intending to make any purchase. But, not to mention that such persons finding on the tables various new publications, are frequently induced, by that opportunity of inspecting them, to become purchasers: this damage does not apply to all booksellers nor to all books; of course it is not necessary to keep in the shop books of small probable demand or great price. In the present case, the retail profit on three copies only, namely, 4*s*. 6*d*., would pay the whole cost of the one copy soiled in the shop; and even that copy might afterwards produce, at an auction, half or a third of its cost price. The argument, therefore, from disappointments in the sale of books, and that arising from heavy stock, are totally groundless in the question between publisher and author. It shold be remarked also, that the publisher is generally a retail, as well as a wholesale, bookseller; and that, besides his profit upon every copy which he sells in his capacity of agent, he is allowed to charge the author as if every copy had been subscribed for at 4*s*. 2*d*., and of course he receives the same profit as the rest of the wholesale traders for the books retailed in his own shop.

394. In the country, there is more reason for a considerable allowance between the retail dealer and the public; because the profit of the country bookseller is diminished by the expense of the carriage / of the books from London. He must also pay a commission, usually five per cent, to his London agent, on all those books which his correspondent does not himself publish. If to this be added a discount of five per cent, allowed for ready money to every customer, and of ten per cent to book clubs, the profit of the bookseller in a small country town is by no means too large.

Some of the writers, who have published criticisms on the observations made in the first edition of this work, have admitted that the apparent rate of profit to the booksellers is *too large*. But they have, on the other hand, urged that too favourable a case is taken in supposing the whole 3000 copies sold. If the reader will turn back to section 382, he will find that the expense of the three first items remains the same, whatever be the number of copies sold; and on looking over the remaining items he will perceive that the bookseller, who incurs very little risk and no outlay, derives exactly the same profit per cent on the copies sold, whatever their numbers may be. This, however, is not the case with the unfortunate author, on whom nearly the whole of the loss falls undivided. The same writers have also maintained, that the profit is fixed at the rate mentioned, in order to enable the bookseller to sustain losses, unavoidably incurred in the purchase and retail of *other* books. This is the weakest of all arguments. It would be equally just that a merchant should charge an extravagant commission for an undertaking unaccompanied with any risk, in order to repay himself for the losses which his own want of skill might lead to in his other mercantile transactions. /

395. That the profit in retailing books is really too large, is proved by several circumstances: First, that the same nominal rate of profit has existed in the bookselling trade for a long series of years, notwithstanding the great fluctuations in the rate of profit on capital invested in every other business. Secondly, that, until very lately, a multitude of booksellers, in all parts of London, were content with a much smaller profit, and were willing to sell for ready money, or at short credit, to persons of undoubted character, at a profit of only ten per cent, and in some instances even at a still smaller percentage, instead of that of twenty-five per cent on the published prices. Thirdly, that they are unable to maintain this rate of profit except by a combination, the object of which is to put down all competition.

396. Some time ago a small number of the large London booksellers entered into such a combination. One of their objects was to prevent any bookseller from selling books for less than ten per cent under the published prices; and in order to enforce this principle, they refuse to sell books, except at the publishing price, to any bookseller who declines signing an agreement to that effect. By degrees, many were prevailed upon to join this combination; and the effect of the exclusion it inflicted, left the small capitalist no option between signing or having his business destroyed. Ultimately, nearly the whole trade, comprising about two thousand four hundred persons, have been compelled to sign the agreement.

As might be naturally expected from a compact so injurious to many of the parties to it, disputes have / arisen; several booksellers have been placed under the ban of the combination, who allege that they have not violated its rules, and who accuse the opposite party of using spies, etc., to entrap them.*

397. The origin of this combination has been explained by Mr Pickering, of Chancery Lane, himself a publisher, in a printed statement, entitled, 'Booksellers' Monopoly' and the following list of booksellers, who form the committee for conducting this combination, is copied from that printed at the head of each of the cases published by Mr Pickering:

Allen, J., 7, Leadenhall Street.
Arch, J., 61, Cornhill.
Baldwin, R., 47, Paternoster Row.
Booth, J.
Duncan, J., 37, Paternoster Row.
Hatchard, J., Piccadilly.
Marshall, R., Stationers' Court.
Murray, J., Albemarle Street.
Rees, O., 39, Paternoster Row.
Richardson, J. M., 23, Cornhill.
Rivington, J., St. Paul's Churchyard.
Wilson, E., Royal Exchange.

* It is now understood that the use of spies has been given up; and it is also known that the system of underselling is again privately resorted to by many; so that the injury arising from this arbitrary system, pursued by the great booksellers, affects only, or most severely, those whose adherence to an extorted promise most deserves respect. *Note to the second edition.*

398. In whatever manner the profits are divided between the publisher and the retail bookseller, the fact remains, that the reader pays for the volume in his hands 6*s*., and that the author will receive only 3*s*. 10*d*.; out of which latter sum, the expense of printing the volume must be paid: so that in passing / through two hands this book has produced a profit of forty-four per cent. This excessive rate of profit has drawn into the book trade a larger share of capital than was really advantageous; and the competition between the different portions of that capital has naturally led to the system of underselling, to which the committee above mentioned are endeavouring to put a stop.*

399. There are two parties who chiefly suffer from this combination – the public and authors. The first party can seldom be induced to take an active part against any grievance; and in fact little is required from it, except a cordial support of the authors, in any attempt to destroy a combination so injurious to the interests of both.

Many an industrious bookseller would be glad to sell for 5*s*. the volume which the reader holds in his hand, and for which he has paid 6*s*.; and, in doing so for *ready money*, the tradesman who paid 4*s*. 6*d*. for the book, would realize, without the least risk, a profit of eleven per cent on the money he had advanced. It is one of the objects of the combination we are discussing, to prevent the small capitalist from employing his capital at that rate of profit which he thinks most advantageous to himself; and such a proceeding is decidedly injurious to the public.

400. Having derived little pecuniary advantage / from my own literary productions; and being aware, that from the very nature of their subjects, they can scarcely be expected to reimburse the expense of preparing them, I may be permitted to offer an opinion upon the subject, which I believe to be as little influenced by any expectation of advantage from the future, as it is by any disappointment at the past.

Before, however, we proceed to sketch the plan of a campaign against Paternoster Row, it will be fit to inform the reader of the

* The monopoly cases, Nos. 1, 2, and 3, of those published by Mr Pickering, should be consulted upon this point; and, as the public will be better able to form a judgement by hearing the other side of the question, it is to be hoped the Chairman of the Committee (Mr Richardson) will *publish* those regulations respecting the trade, a copy of which, Mr Pickering states, is refused by the Committee *even to those who sign them.*

nature of the enemies' forces, and of his means of attack and defence. Several of the great publishers find it convenient to be the proprietors of *reviews*, *magazines*, *journals*, and even of *newspapers*. The *editors* are paid, in some instances very handsomely, for their superintendence; and it is scarcely to be expected that they should always mete out the severest justice on works by the sale of which their employers are enriched. The great and popular works of the day are, of course, reviewed with some care, and with deference to public opinion. Without this, the journals would not sell; and it is convenient to be able to quote such articles as instances of impartiality. Under shelter of this, a host of ephemeral productions are written into a transitory popularity; and by the aid of this process, the shelves of the booksellers, as well as the pockets of the public, are disencumbered. To such an extent are these means employed, that some of the periodical publications of the day ought to be regarded merely as *advertising machines*. That the reader may be in some measure on his guard against such modes of influencing his judgement, he should examine whether the work reviewed is published by / the bookseller who is the proprietor of the review; a fact which can sometimes be ascertained from the title of the book as given at the head of the article. But this is by no means a certain criterion, because partnerships in various publications exist between houses in the book trade, which are not generally known to the public; so that, in fact, until reviews are established in which booksellers have no interest, they can never be safely trusted.

401. In order to put down the combination of booksellers, no plan appears so likely to succeed as a counter-association of authors. If any considerable portion of the literary world were to unite and form such an association; and if its affairs were directed by an active committee, much might be accomplished. The objects of such an union should be, to employ some person well skilled in the printing, and in the bookselling trade; and to establish him in some central situation as their agent. Each member of the association to be at liberty to place any, or all of his works in the hands of this agent for sale; to allow any advertisements, or list of books published by members of the association, to be stitched up at the end of each of his own productions; the expense of preparing them being defrayed by the proprietors of the books advertised.

The duties of the agent would be to retail to the public, for *ready money*, copies of books published by members of the association. To sell to the trade, at prices agreed upon, any copies they may require. To cause to be inserted in the journals, or at the end of works published by members, any advertisements which the committee or authors may / direct. To prepare a general catalogue of the works of members. To be the agent for any member of the association respecting the printing of any work.

Such a union would naturally present other advantages; and as each author would retain the liberty of putting any price he might think fit on his productions, the public would have the advantage of reduction in price produced by competition between authors on the same subject, as well as of that arising from a cheaper mode of publishing the volumes sold to them.

402. Possibly, one of the consequences resulting from such an association, would be the establishment of a good and an impartial review, a work the want of which has been felt for several years. The two long-established and celebrated reviews, the unbending champions of the most opposite political opinions, are, from widely differing causes, exhibiting unequivocal signs of decrepitude and decay. The quarterly advocate of despotic principles is fast receding from the advancing intelligence of the age; the new strength and new position which that intelligence has acquired, demands for its expression, new organs, equally the representatives of its intellectual power, and of its moral energies: whilst, on the other hand, the sceptre of the northern critics has passed, from the vigorous grasp of those who established its dominion, into feebler hands.

403. It may be stated as a difficulty in realizing this suggestion, that those most competent to supply periodical criticism, are already engaged. But it is to be observed, that there are many who now supply literary criticisms to journals, the political principles / of which they disapprove; and that if once a respectable and well-supported review* were established, capable of competing, in payment to its contributors, with the wealthiest of its rivals, it would very soon be supplied with the best materials the country can produce.† It may

* At the moment when this opinion as to the necessity for a new review was passing through the press, I was informed that the elements of such an undertaking were already organized.

† I has been suggested to me, that the doctrines maintained in this chapter may subject the present volume to the opposition of that combination which it has opposed. I do not entertain that opinion; and for this reason, that the booksellers are too shrewd a class to supply such an admirable passport to publicity as their opposition would

also be apprehended that such a combination of authors would be favourable to each other. There are two temptations to which an editor of a review is commonly exposed: the first is, a tendency to consult too much, in the works he criticizes, the interest of the proprietor of his review; the second, a similar inclination to consult the interests of his friends. The plan which has been proposed removes one of these temptations, but it would be very difficult, if not impossible, to destroy the other. /

prove to be if generally suspected.[1] But should my readers take a different view of the question, they can easily assist in remedying the evil, by each mentioning the existence of this little volume to two of his friends.

[1] I was mistaken in this conjecture; *all* booksellers are not so shrewd as I had imagined, for some did refuse to sell this volume; consequently others sold a larger number of copies.

In the preface to the second edition, at the commencement of this volume, the reader will find some further observations on the effect of the booksellers' combination. *Note to the Second Edition.*

[334/5]

CHAPTER XXXII

ON THE EFFECT OF MACHINERY IN REDUCING THE DEMAND FOR LABOUR

404. One of the objections most frequently urged against machinery is, that it has a tendency to supersede much of the hand labour which was previously employed; and in fact unless a machine diminished the labour necessary to make an article, it could never come into use. But if it have that effect, its owner, in order to extend the sale of his produce, will be obliged to undersell his competitors; this will induce them also to introduce the new machine, and the effect of this competition will soon cause the article to fall, until the profits on capital, under the new system, shall be reduced to the same rate as under the old. Although, therefore, the use of machinery has at first a tendency to throw labour out of employment, yet the increased demand consequent upon the reduced price, almost immediately absorbs a considerable portion of that labour, and perhaps, in some cases, the whole of what would otherwise have been displaced.

That the effect of a new machine is to diminish the labour required for the production of the *same* quantity of manufactured commodities may be clearly perceived, by imagining a society, in which occupation are not divided, each man himself manufacturing all the articles he consumes. Supposing each / individual to labour during ten hours daily, one of which is devoted to making shoes, it is evident that if any tool or machine be introduced, by the use of which his shoes can be made in half the usual time, then each member of the community will enjoy the same comforts as before by only nine and one-half hours' labour.

405. If, therefore, we wish to prove that the total quantity of labour is not diminished by the introduction of machines, we must have recourse to some other principle of our nature. But the same motive which urges a man to activity will become additionally powerful, when he finds his comforts procured with diminished labour; and in such circumstances, it is probable, that many would employ the time thus redeemed in

contriving new tools for other branches of their occupations. He who has habitually worked ten hours a day, will employ the half hour saved by the new machine in gratifying some other want; and as each new machine adds to these gratifications, new luxuries will open to his view, which continued enjoyment will as surely render necessary to his happiness.

406. In countries where occupations are divided, and where the division of labour is practised, the ultimate consequence of improvements in machinery is almost invariably to cause a greater demand for labour. Frequently the new labour requires, at its commencement, a higher degree of skill than the old; and, unfortunately, the class of persons driven out of the old employment are not always qualified for the new one; so that a certain interval must elapse before the whole of their labour is wanted. / This, for a time, produces considerable suffering amongst the working classes; and it is of great importance for their happiness that they should be aware of these effects, and bc enabled to foresee them at an early period, in order to diminish, as much as possible, the injury resulting from them.

407. One very important enquiry which this subject presents is the question – *whether it is more for the interest of the working classes, that improved machinery should be so perfect as to defy the competition of hand labour; and that they should thus be at once driven out of the trade by it; or be gradually forced to quit it by the slow and successive advances of the machine?* The suffering which arises from a quick transition is undoubtedly more intense; but it is also much less permanent than that which results from the slower process: and if the competition is perceived to be perfectly hopeless, the workman will at once set himself to learn a new department of his art. On the other hand, although new machinery causes an increased demand for skill in those who make and repair it, and in those who first superintend its use; yet there are other cases in which it enables children and inferior workmen to execute work that previously required greater skill. In such circumstances, even though the increased demand for the article, produced by its diminished price, should speedily give occupation to all who were before employed, yet the very diminution of the skill required, would open a wider field of competition amongst the working classes themselves.

That machines do not, even at their first introduction, *invariably* throw human labour out of / employment, must be admitted; and it has been maintained, by persons very competent to form an opinion on the subject, that they never produce that effect. The solution of this question depends on facts, which, unfortunately, have not yet been collected; and the circumstance of our not possessing the data necessary for the full examination of so important a subject, supplies an additional reason for impressing, upon the minds of all who are interested in such enquiries, the importance of procuring accurate registries, at various times, of the number of persons employed in particular branches of manufacture, of the number of machines used by them, and of the wages they receive.

408. In relation to the enquiry just mentioned, I shall offer some remarks upon the facts within my knowledge, and only regret that those which I can support by numerical statement are so few. When the *crushing mill*, used in Cornwall and other mining countries, superseded the labour of a great number of young women, who worked very hard in breaking ores with flat hammers, no distress followed. The reason of this appears to have been, that the proprietors of the mines, having one portion of their capital released by the superior cheapness of the process executed by the mills, found it their interest to apply more labour to other operations. The women, disengaged from mere drudgery, were thus profitably employed in *dressing* the ores, a work which required skill and judgement in the selection.

409. The increased production arising from alterations in the machinery, or from improved modes of using it, appears from the following table. A machine / called in the cotton manufacture a 'stretcher', worked by one man, produced as follows:

Year	*Pounds of cotton spun*	*Roving wages per score*	*Rate of earning per week*
		s. d.	*s. d.*
1810	400	1 3½	25 10*
1811	600	0 10	25 0
1813	850	0 9	31 10½
1823	1000	0 7½	31 3

The same man working at another stretcher, the roving a little finer, produced,

1823	900	0 7½	28 1½
1825	1000	0 7	27 6
1827	1200	0 6	30 0
1832	1200	0 6	30 0

* In 1810, the workman's wages were guaranteed not to be less than 26*s*.

In this instance, production has gradually increased until, at the end of twenty-two years, three times as much work is done as at the commencement, although the manual labour employed remains the same. The weekly earnings of the workmen have not fluctuated very much, and appear, on the whole, to have advanced: but it would be imprudent to push too far reasonings founded upon a single instance.

410. The produce of 480 spindles of 'mule yarn spinning', at different periods, was as follows:

Year	*Hanks about 40 to the pound*	*Wages per thousand*
		s. d.
1806	6668	9 2
1823	8000	6 3
1832	10,000	3 8

411. The subjoined view of the state of weaving by hand- and by power-looms, at Stockport, in the / years 1822 and 1832, is taken from an enumeration of the machines contained in 65 factories, and was collected for the purpose of being given in evidence before a Committee of the House of Commons.

	In 1822	*In 1832*	
Hand-loom weavers	2800	800	2000 decrease
Persons using power-looms	657	3059	2402 increase
Persons to dress the warp	98	388	290 increase
Total persons employed	3555	4247	692 increase
Power-looms	1970	9177	8207 increase

During this period, the number of hand-looms in employment has diminished to less than one-third, whilst that of power-looms has increased to more than five times its former amount. The total number of workmen has increased about one-third; but the amount of manufactured goods (supposing each power-loom to do only the work of three hand-looms) is three and a half times as large as it was before.

412. In considering this increase of employment, it must be admitted, that the two thousand persons thrown out of work are not exactly of the same class as those called into employment by the power-looms. A hand-weaver must possess bodily strength, which is not essential for a person attending a power-loom; consequently, women and young persons of both sexes, from fifteen to seventeen years of age,

find employment in power-loom factories. This, however, would be a very limited view of the employment arising from the introduction of power-looms: the skill called into action in building the new factories, in constructing the new machinery, in making the / steam-engines to drive it, and in devising improvements in the structure of the looms, as well as in regulating the economy of the establishment, is of a much higher order than that which it had assisted in superseding; and if we possessed any means of measuring this, it would probably be found larger in amount. Nor, in this view of the subject, must we omit the fact, that although hand-looms would have increased in number if those moved by steam had not been invented, yet it is the cheapness of the article manufactured by power-looms which has caused this great extension of their employment, and that by diminishing the price of one article of commerce, we always call into additional activity the energy of those who produce others. It appears that the number of hand-looms in use in England and Scotland in 1830, was about 240,000; nearly the same number existed in the year 1820: whereas the number of power-looms which, in 1830, was 55,000, had, in 1820, been 14,000. When it is considered that each of these power-looms did as much work as three worked by hand, the increased producing power was equal to that of 123,000 hand-looms. During the whole of this period the wages and employment of hand-loom weavers have been very precarious.

413. Increased intelligence amongst the working classes, may enable them to foresee some of those improvements which are likely for a time to affect the value of their labour; and the assistance of savings banks and friendly societies, (the advantages of which can never be too frequently, or too strongly, pressed upon their attention), may be / of some avail in remedying the evil: but it may be useful also to suggest to them, that a diversity of employments amongst the members of one family will tend, in some measure, to mitigate the privations which arise from fluctuation in the value of labour. /

[342/3]

CHAPTER XXXIII

ON THE EFFECT OF TAXES AND OF LEGAL RESTRICTIONS UPON MANUFACTURES

414. As soon as a tax is put upon any article, the ingenuity of those who make, and of those who use it, is directed to the means of evading as large a part of the tax as they can; and this may often be accomplished in ways which are perfectly fair and legal. An excise duty exists at present of 3*d*.* per pound upon all writing paper. The effect of this impost is, that much of the paper which is employed, is made extremely thin, in order that the weight of a given number of sheets may be as small as possible. Soon after the first imposition of the tax upon windows, which depended upon their number, and not upon their size, new-built houses began to have fewer windows and those of larger dimensions than before. Staircases were lighted by extremely long windows, illuminating three or four flights of stairs. When the tax was increased, and the *size* of windows charged as single was limited, then still greater care was taken to have as few windows as possible, and internal lights became frequent. These internal lights in their turn became the subject of taxation; but it was easy to evade the discovery of them, and / in the last Act of Parliament reducing the assessed taxes, they ceased to be chargeable. From the changes thus successively introduced in the number the forms, and the positions of the windows, a tolerable conjecture might, in some instances, be formed of the age of a house.

415. A tax on windows is exposed to objection on the double ground of its excluding *air* and *light*, and it is on both accounts injurious to health. The importance of *light* to the enjoyment of health is not perhaps sufficiently appreciated: in the cold and more variable climates, it is of still greater importance than in warmer countries.

* Twenty-eight shillings per cwt for the finer, twenty-one shillings per cwt for the coarser papers.

416. The effects of regulations of excise upon our home manufactures are often productive of great inconvenience; and check, materially, the natural progress of improvement. It is frequently necessary, for the purposes of revenue, to oblige manufacturers to take out a license, and to compel them to work according to certain rules, and to make certain stated quantities at each operation. When these quantities are large, as in general they are, they deter manufacturers from making experiments, and thus impede improvements both in the mode of conducting the processes and in the introduction of new materials. Difficulties of this nature have occurred in experimenting upon glass for optical purposes; but in this case, permission has been obtained by fit persons to make experiments, without the interference of the excise. It ought, however, to be remembered, that such permission, if frequently or indiscriminately granted, might be abused: the greatest protection against such an abuse will be found, in bringing the / force of public opinion to bear upon scientific men – and thus enabling the proper authorities, although themselves but moderately conversant with science, to judge of the propriety of the permission, from the public character of the applicant.

417. From the evidence given, in 1808, before the Committee of the House of Commons, *On Distillation from Sugar and Molasses*, it appeared that, by a different mode of working from that prescribed by the Excise, the spirits from a given weight of corn, which then produced eighteen gallons, might easily have been increased to twenty gallons. Nothing more is required for this purpose, than to make what is called the *wash* weaker; the consequence of which is, that fermentation goes on to a greater extent. It was stated, however, that such a deviation would render the collection of the duty liable to great difficulties; and that it would not benefit the distiller much, since his price was enhanced to the customer by any increase of expense in the fabrication. Here then is a case in which a quantity, amounting to one-ninth of the total produce, is actually lost to the country. A similar effect arises in the coal trade, from the effect of a duty; for, according to the evidence before the House of Commons, it appears that a considerable quantity of the very best coal is actually wasted. The extent of this waste is very various in different mines; but in some cases it amounts to one-third.

418. The effects of duties upon the import of foreign manufactures are equally curious. A singular instance occurred in the United States, where *bar-iron* was, on its introduction, liable to a duty of /

140 per cent *ad valorem*, whilst *hardware* was charged at 25 per cent only. In consequence of this tax, large quantities of malleable iron rails for railroads were imported into America under the denomination of hardware; the difference of 115 per cent in duty more than counter balancing the expense of fashioning the iron into rails prior to its importation.

419. Duties, drawbacks, and bounties, when considerable in amount, are all liable to objections of a very serious nature, from the frauds to which they give rise. It has been stated before Committees of the House of Commons, that calicoes made up in the form, and with the appearance of linen, have frequently been exported for the purpose of obtaining the bounty: for calico made up in this way sells only at 1*s*. 4*d*. per yard, whereas linen of equal fineness is worth from 2*s*. 8*d*. to 2*s*. 10*d*. per yard. It appeared from the evidence, that one house in six months sold five hundred such pieces of calico.

In almost all cases heavy duties, or prohibitions, are ineffective as well as injurious; for unless the articles excluded are of very large dimensions, there constantly arises a price at which they will be clandestinely imported by the smuggler. The extent, therefore, to which smuggling can be carried, should always be considered in the imposition of new duties, or in the alteration of old ones. Unfortunately it has been pushed so far, and is so systematically conducted between this country and France, that the price per cent at which most contraband articles can be procured is perfectly well known. From the evidence of Mr Galloway, it appears that, from 30 / to 40 per cent was the rate of insurance on exporting prohibited machinery from England, and that the larger the quantity the less was the percentage demanded. From evidence given in the Report of the Watch and Clock-makers' Committee, in 1817, it appears that persons were constantly in the habit of receiving in France watches, lace, silks, and other articles of value easily portable, and delivering them in England at ten per cent on their estimated worth, in which sum the cost of transport and the risk of smuggling were included.

420. The process employed in manufacturing often depends upon the mode in which a tax is levied on the materials, or on the article produced. Watch glasses are made in England by workmen who purchase from the glass house globes of five or six inches in diameter, out of which, by means of a piece of red-hot tobacco pipe, guided round a pattern watch glass placed on the globe, they crack five others: these are afterwards ground and smoothed on the edges. In

the Tyrol the rough watch glasses are supplied at once from the glass house; the workman, applying a thick ring of cold glass to each globe as soon as it is blown, causes a piece, of the size of a watch glass, to be cracked out. The remaining portion of the globe is immediately broken, and returns to the melting pot. This process could not be adopted in England with the same economy, because the whole of the glass taken out of the pot is subject to the excise duty.

421. The objections thus stated as incidental to particular modes of taxation are not raised with a view to the removal of those particular taxes; / their fitness or unfitness must be decided by a much wider enquiry, into which it is not the object of this volume to enter. Taxes are essential for the security both of liberty and property, and the evils which have been mentioned may be the least amongst those which might have been chosen. It is, however, important that the various effects of every tax should be studied, and that those should be adopted which, upon the whole, are found to give the least check to the productive industry of the country.

422. In enquiring into the effect produced, or to be apprehended from any particular mode of taxation, it is necessary to examine a little into the interests of the parties who approve of the plan in question, as well as of those who object to it. Instances have occurred where the persons paying a tax into the hands of government have themselves been adverse to any reduction. This happened in the case of one class of calico-printers, whose interest really was injured by a removal of the tax on the printing: they received from the manufacturers, payment for the duty, about two months before they were themselves called on to pay it to government; and the consequence was, that a considerable capital always remained in their hands. The evidence which states this circumstance is well calculated to promote a reasonable circumspection in such enquiries.

Question. Do you happen to know anything of an opposition from calico-printers to the repeal of the tax on printed calicoes?

Answer. I have certainly heard of such an opposition, and am not surprised at it. There are very few individuals / who are, in fact, interested in the non-repeal of the tax; there are two classes of calico-printers; one, who print their own cloth, send their goods into the market, and sell them on their own account; they frequently advance the duty to government, and pay it in cash before their goods are sold, but generally before the goods are paid for, being most commonly sold on a credit of six months: they are of course interested on that account, as well as on others that have been stated, in the repeal of the tax. The other class of calico-printers print the cloth of other people; they print for hire, and on re-delivery of the cloth when printed, they receive the amount of the duty, which they are not called upon to pay to government sooner, on an average, than nine weeks from the stamping of the goods. Where the business is carried on upon a large scale, the arrears of duty due to government often amount to eight, or even ten thousand pounds, and furnish a capital with which these gentlemen carry on their business; it is not, therefore, to be wondered at that they should be opposed to the prayer of our petition.

423. The policy of giving bounties upon home productions, and of enforcing restrictions against those which can be produced more cheaply in other countries, is of a very questionable nature: and, except for the purpose of introducing a new manufacture, in a country where there is not much commercial or manufacturing spirit, is scarcely to be defended. All incidental modes of taxing one class of the community, the *consumers*, to an unknown extent, for the sake of supporting another class, the *manufacturers*, who would otherwise abandon that mode of employing their capital, are highly objectionable. One part of the price of any article produced under such circumstances, consists of / the expenditure, together with the ordinary profits of capital: the other part of its price may be looked upon as charity, given to induce the manufacturer to continue an unprofitable use of his capital, in order to give employment to his workmen. If the sum of what the consumers are thus forced to pay, merely on account of these artificial restrictions, where generally known, its amount would astonish *even those who advocate them*; and it would be evident to both parties, that the employment of capital in those branches of trade ought to be abandoned.

424. The restriction of articles produced in a manufactory to certain sizes, is attended with some good effect in an economical view, arising chiefly from the smaller number of different tools required in making them, as well as from less frequent change in the adjustment of those tools. A similar source of economy is employed in the Navy: by having ships divided into a certain number of classes, each of which comprises vessels of the same dimensions, the rigging made for one vessel will fit any other of its class; a circumstance which renders the supply of distant stations more easy.

425. The effects of the removal of a monopoly are often very important, and they were perhaps never more remarkable than in the bobbin net trade, in the years 1824 and 1825. These effects were,

however, considerably enhanced by the general rage for speculation which was so prevalent during that singular period. One of the patents of Mr Heathcote for a bobbin net machine had just then expired, whilst another, for an improvement in a particular part of such machines, called a *turn again*, had yet a few / years to run. Many licenses had been granted to use the former patent, which were charged at the rate of about five pounds per annum for each quarter of a yard in width, so that what is termed *a six-quarter frame* (which makes bobbin net a yard and a half wide) paid thirty pounds a year. The second patent was ultimately abandoned in August, 1823, infringements of it having taken place.

It was not surprising that, on the removal of the monopoly arising from this patent, a multitude of persons became desirous of embarking in a trade which had hitherto yielded a very large profit. The bobbin net machine occupies little space; and is, from that circumstance, well adapted for a domestic manufacture. The machines which already existed, were principally in the hands of the manufacturers; but, a kind of mania for obtaining them seized on persons of all descriptions, who could raise a small capital; and, under its influence, butchers, bakers, small farmers, publicans, gentlemen's servants, and, in some cases, even clergymen, became anxious to possess bobbin net machines.

Some few machines were rented; but, in most of these cases, the workman purchased the machine he employed, by instalments of from £3 to £6 weekly, for a *six quarter machine*; and many individuals, unacquainted with the mode of using the machines so purchased, paid others of more experience for instructing them in their use; £50 or £60 being sometimes given for this instruction. The success of the first speculators induced others to follow the example; and the machine-makers were almost overwhelmed with orders for lace frames. Such was the / desire to procure them, that many persons deposited a large part, or the whole, of the price, in the hands of the frame-makers, in order to insure their having the earliest supply. This, as might naturally be expected, raised the price of wages amongst the workmen employed in machine-making; and the effect was felt at a considerable distance from Nottingham, which was the centre of this mania. Smiths not used to *flat filing*, coming from distant parts, earned from 30*s*. to 42*s*. per week. Finishing smiths, accustomed to the work, gained from £3 to £4 per week. The forging smith, if accustomed to his work, gained from £5 to £6 per week, and some few earned £10 per week. In making what are technically called *insides*, those who were best paid, were generally clock- and watch-makers, from all the districts round, who received from £3 to £4 per week. The *setters-up* – persons who put the parts of the machine together – charged £20 for their assistance; and, a *six quarter machine*, could be put together in a fortnight or three weeks.

426. Good workmen, being thus induced to desert less profitable branches of their business, in order to supply this extraordinary demand, the masters, in other trades, soon found their men leaving them, without being aware of the immediate reason: some of the more intelligent, however, ascertained the cause. They went from Birmingham to Nottingham, in order to examine into the circumstances which had seduced almost all the journeymen clockmakers from their own workshops; and it was soon apparent, that the men who had been working as clockmakers in Birmingham, at the rate of 25*s*. a week, could earn / £2 by working at lace frame-making in Nottingham.

On examining the nature of this profitable work, the master clockmakers perceived that one part of the bobbin net machines, that which held the bobbins, could easily be made in their own workshops. They therefore contracted with the machine-makers, who had already more work ordered than they could execute, to supply the *bobbin carriers*, at a price which enabled them, on their return home, to give such increased wages as were sufficient to retain their own workmen, as well as yield themselves a good profit. Thus an additional facility was afforded for the construction of these bobbin net machines: and the conclusion was not difficult to be foreseen. The immense supply of bobbin net thus poured into the market, speedily reduced its price; this reduction in price, rendered the machines by which the net was made, less valuable; some few of the earliest producers, for a short time, carried on a profitable trade; but multitudes were disappointed, and many ruined. The low price at which the fabric sold, together with its lightness and beauty, combined to extend the sale; and ultimately, new improvements in the machines, rendered the older ones still less valuable.

427. The bobbin net trade is, at present, both extensive and increasing; and, as it may, probably, claim a larger portion of public attention at some future time, it will be interesting to describe briefly its actual state.

A lace frame on the most improved principle, at the present day, manufacturing a piece of net / two yards wide, when worked night and day, will produce six hundred and twenty *racks* per week. A *rack* is two hundred and forty holes; and as in the machine to which we refer, three *racks* are equal in length to one yard, it will produce 21,493 square yards of bobbin net annually. Three men keep this machine constantly working; and, they were paid (by piece-work) about 25*s*. each per week, in 1830. Two boys, working only in the day-time, can prepare the bobbins for this machine, and are paid from 2*s*. to 4*s*. per week, according to their skill. Forty-six square yards of this net weigh two pounds three ounces; so that each square yard weighs a little more than three-quarters of an ounce.

428. For a condensed and general view of the present state of this trade, we shall avail ourselves of a statement by Mr William Felkin, of Nottingham, dated September, 1831, and entitled *Facts and Calculations illustrative of the Present State of the Bobbin Net Trade*. It appears to have been collected with care, and contains, in a single sheet of paper, a body of facts of the greatest importance.*

429. The total capital employed in the factories, for preparing the cotton, in those for weaving the bobbin net, and in various processes to which it is subject, is estimated at above £2,000,000, and the number of persons who receive wages, at above two hundred thousand. /

Comparison of the value of the raw material imported, with the value of the goods manufactured therefrom

Amount of Sea Island cotton annually used 1,600,000 lbs., value £120,000; this is manufactured into yarn, weighing 1,000,000 lbs., value £500,000.

There is also used 25,000 lbs. of raw silk, which costs £30,000, and is doubled into 20,000 lbs. thrown, worth £40,000.

Raw Material	*Manufacture*	*Square yards produced*	*Value per sq. yd*	*Total value*
			s. d.	*£*
Cotton, 1,600,000 *lbs*	Power Net	6,750,000	1 3	421,875
	Hand ditto	15,750,000	1 9	1,378,125
	Fancy ditto	150,000	3 6	26,250
Silk, 25,000 *lbs*	Silk Goods	750,000	1 9	65,625
		23,400,000		1,891,875

* I cannot omit the opportunity of expressing my hope that this example will be followed in other trades. We should thus obtain a body of information equally important to the workman, the capitalist, the philosopher, and the statesman.

The brown nets which are sold in the Nottingham market, are in part disposed of by the agents of twelve or fifteen of the larger makers, i.e. to the amount of about £250,000 a year. The principal part of the remainder, i.e. about £1,050,000 a year, is sold by about two hundred agents, who take the goods from one warehouse to another for sale.

Of this production, about half is exported in the unembroidered state. The exports of bobbin net are in great part to Hamburgh, for sale at home and at Leipzic and Frankfort fairs, Antwerp, and the rest of Belgium; to France, by / contraband; to Italy, and North and South America. Though a very suitable article, yet the quantity sent eastward of the Cape of Good Hope, has hitherto been too trifling for notice. Three-eighths of the whole production are sold unembroidered at home. The remaining one-eighth is embroidered in this country, and increases the ultimate value as under, viz.

Embroidery	*Increases value*	*Ultimate worth*
	£	£
On power net	131,840	553,715
On hand net	1,205,860	2,583,985
On fancy net	78,750	105,000
On silk net	109,375	175,000
	Total embroidery, wages, and profit, 1,525,825	Ultimate total value 3,417,700

From this it appears, that in the operations of this trade, which had no existence twenty years ago, £120,000 original cost of cotton becomes, when manufactured, of the ultimate value of £3,242,700 sterling.

As to weekly wages paid, I hazard the following as the judgement of those conversant with the respective branches, viz.

In fine spinning and doubling, adults 25*s*.; children 7*s*.: work twelve hours per day.

In bobbin net making; men working machines, 18*s*.; apprentices, youths of fifteen or more, 10*s*.; by power, fifteen hours; by hand, eight to twelve hours, according to width.

In mending; children 4*s*.: women 8*s*.; work nine to fourteen hours, *ad libitum*. /

In winding, threading, etc., children and young women, 5*s*.; irregular work, according to the progress of machines.

In embroidery; children seven years old and upwards, 1*s*. to 3*s*.; work ten to twelve hours; women, if regularly at work, 5*s*. to 7*s*. 6*d*.; twelve to fourteen hours.

As an example of the effect of the wages of lace embroidery, etc., it may be observed, it is often the case that a stocking weaver in a country village will earn only 7*s*. a week, and his wife and children 7*s*. to 14*s*. more at the embroidery frame.

430. The principal part of the hand-machines employed in the bobbin net manufacture are worked in shops, forming part of, or attached to, private houses. The subjoined list will show the kinds of machinery employed, and classes of persons to whom it belongs.

Bobbin net machinery now at work in the Kingdom

Hand levers	6 quarter	500	Hand circulars	6 quarter	100
	7 quarter	200		7 quarter	300
	8 quarter	300		8 quarter	400
	10 quarter	300		9 quarter	100
	12 quarter	30		10 quarter	300
	16 quarter	20		12 quarter	100
	20 quarter	1	Hand transverse, pusher,		
Hand rotary	10 quarter	50	straight bolt, etc. averaging	5 quarters	750
	12 quarter	50			——
		——			2050
		1451			

	Total hand machines	3501 /
Power	6 quarter	100
	7 quarter	40
	8 quarter	350
	10 quarter	270
	12 quarter	220
	16 quarter	20
	Total power machines	1000
		——
	Total number of machines	4501

700 persons own	1 machine,	700 machines.
226	2	452
181	3	543
96	4	384
40	5	200
21	6	126
17	7	119
19	8	152
17	9	153
12	10	120
8	11	88
6	12	72
5	13	65

5 persons own 14 machines,		70 machines.
4	16	64
25 own respectively 18, 19, 20, 21, 23, 24, 25, 26, 27, 28, 29, 30, 32, 33, 35, 36, 37, 50, 60, 68, 70, 75, 95, 105, 206		1192

Number of owners of machines } 1382 — Holding together 4500 machines.

The hand workmen consist of the above-named owners	1000
And of journeymen and apprentices	4000
	5000 /

These machines are distributed as follows

Nottingham	1240
New Radford	140
Old Radford and Bloomsgrove	240
Ison Green	160
Beeston and Chilwell	130
New and Old Snenton	180
Derby and its vicinity	185
Loughborough and its vicinity	385
Leicester	95
Mansfield	85
Tiverton	220
Barnstable	180
Chard	190
Isle of Wight	80
In sundry other places	990
	4500

Of the above owners, one thousand work in their own machines, and enter into the class of journeymen as well as that of masters in operating on the rate of wages. If they reduce the price of their goods in the market, they reduce their own wages first; and, of course, eventually the rate of wages throughout the trade. It is a very lamentable fact, that one-half, or more, of the one thousand one hundred persons specified in the list as owning one, two, and three machines, have been compelled to mortgage their machines for more than their worth in the market, and are in many cases totally insolvent. Their machines are principally narrow and making short pieces, while the absurd system of bleaching at so much a piece goods of all lengths and widths, and dressing at / so much all widths, has caused the new machines to be all wide, and capable of producing long pieces; of course to the serious disadvantage, if not utter ruin, of the small owner of narrow machines.

It has been observed above, that wages have been reduced, say 25 per cent in the last two years, or from 24*s*. to 18*s*. a week. Machines have increased in the same time one-eighth in number, or from four thousand to four thousand five hundred, and one-sixth in capacity of production. It is deserving the serious notice of all proprietors of existing machines, that machines are now introducing into the trade of such power of production as must still more than ever depreciate (in the absence of an immensely increased demand) the value of their property.

431. From this abstract, we may form some judgement of the importance of the bobbin net trade. But the extent to which it bids fair to be carried in future, when the eastern markets shall be more open to our industry, may be conjectured from the fact which Mr Felkin subsequently states – that 'We can export a durable and elegant article in cotton bobbin net, at 4*d*. per square yard, proper for certain useful and ornamental purposes, as curtains, etc.; and another article used for many purposes in female dress at 6*d*. the square yard.'

432. *Of patents.* In order to encourage the invention, the improvement, or the importation of machines, and of discoveries relating to manufactures, it has been the practice in many countries, to grant to the inventors or first introducers, an exclusive / privilege for a term of years. Such monopolies are termed patents; and they are granted, on the payment of certain fees, for different periods, from five to twenty years.

The following table, compiled from the Report of the Committee of the House of Commons *on Patents*, 1829, shows the expense and duration of patents in various countries:

Countries	*Expense* £ *s*. *d*.	*Term of years*	*Number granted in six years, ending in 1826. – (*Rep. *p. 243.)*
England	120 0 0	14	914
Ireland	125 0 0	14	
Scotland	100 0 0	14	
America	6 15 0	14	

Countries	*Expense*	*Term of years*	*Number granted in six years, ending in 1826. – (Rep. p. 243.)*
	£ *s.* *d.*		
France	12 0 0	5	1091
	32 0 0	10	
	60 0 0	15	
Netherlands	£6 to £30	5, 10, 15	
Austria	42 10 0	15	1099
Spain* – Inventor	20 9 4	15	
Improver	12 5 7	10	
Importer	10 4 8	6	

433. It is clearly of importance to preserve to each inventor the sole use of his invention, until he / shall have been amply repaid for the risk and expense to which he has been exposed, as well as for the talent he has exerted in completing it. But, the degrees of merit are so various, and the difficulties of legislating upon the subject so great, that it has been found almost impossible to frame a law which shall not, practically, be open to the most serious objections.

The difficulty of defending an English patent in any judicial trial, is very great; and the number of instances on record in which the defence has succeeded, are comparatively few. This circumstance has induced some manufacturers, no longer to regard a patent as a privilege by which a monopoly price may be secured: but they sell the patent article at such a price, as will merely produce the ordinary profits of capital; and thus secure to themselves the fabrication of it, because no competitors can derive a profit from invading a patent so exercised.

434. The law of copyright, is, in some measure, allied to that of patents; and it is curious to observe, that those species of property which require the highest talent, and the greatest cultivation – which are, more than any other, the pure creations of mind – should have been the latest to be recognized by the State. Fortunately, the means of deciding on an infringement of property in regard to a literary production, are not very difficult; but the present laws are, in some cases, productive of considerable hardship, as well as of impediment to the advancement of knowledge.

* The expense of a patent in Spain is stated in the report to be respectively 2000, 1200, and 1000 reals. If these are reals of *vellon*, in which accounts are usually kept at Madrid, the above sums are correct; but if they are reals of *plate*, the above sums ought to be nearly doubled.

435. Whilst discussing the general expediency of limitations and restrictions, it may be desirable to / point out one which seems to promise advantage, though by no means free from grave objections. The question of permitting by law, the existence of partnerships in which the responsibility of one or more of the partners is limited in amount, is peculiarly important in a manufacturing, as well as a commercial point of view. In the former light, it appears calculated to aid that division of labour, which we have already proved to be as advantageous in mental as it is in bodily operations; and it might possibly give rise to a more advantageous distribution of talent, and its combinations, than at present exists. There are in this country, many persons possessed of moderate capital, who do not themselves enjoy the power of invention in the mechanical and chemical arts, but who are tolerable judges of such inventions, and excellent judges of human character. Such persons might, with great success, employ themselves in finding out inventive workmen, whose want of capital prevents them from realizing their projects. If they could enter into a limited partnership with persons so circumstanced, they might restrain within proper bounds the imagination of the inventor, and by supplying capital to judicious schemes, render a service to the country, and secure a profit for themselves.

436. Amongst the restrictions intended for the general benefit of our manufacturers, there existed a few years ago one by which workmen were forbidden to go out of the country. A law so completely at variance with every principle of liberty, ought never to have been enacted. It was not, however, until experience had convinced the legislature / of its inefficiency, that it was repealed.* When, after the last war, the renewed intercourse between England and the Continent became extensive, it was soon found that it was impossible to discover the various disguises which the workmen could assume; and the effect of the law was rather, by the fear of punishment, to deter those who had left the country from returning, than to check their disposition to migrate.

436.* The principle, *that government ought to interfere as little as possible between workmen and their employers*, is so well established, that it is important to guard against its misapplication. It is not

* In the year 1824 the law against workmen going abroad, as well as the laws preventing them from combining, were repealed, after the fullest enquiry by a Committee of the House of Commons. In 1825 an attempt to re-enact some of the most objectionable was made, but it failed.

inconsistent with this principle to insist on the workmen being paid in money – for this is merely to protect them from being deceived; and still less is it a deviation from it to limit the number of hours during which children shall work in factories, or the age at which they shall commence that species of labour – for they are not free agents, nor are they capable of judging, if they were; and both policy and humanity concur in demanding for them some legislative protection. In both cases it is as right and politic to protect the weaker party from fraud or force, as it would be impolitic and unjust to interfere with the *amount* of the wages of either. /

[364/5]

CHAPTER XXXIV

ON THE EXPORTATION OF MACHINERY

437. A few years only have elapsed, since our workmen were not merely prohibited by Act of Parliament from transporting themselves to countries in which their industry would produce for them higher wages, but were forbidden to export the greater part of the machinery which they were employed to manufacture at home. The reason assigned for this prohibition was, the apprehension that foreigners might avail themselves of our improved machinery, and thus compete with our manufacturers. It was, in fact, a sacrifice of the interests of one class of persons, the makers of machinery, for the imagined benefit of another class, those who use it. Now, independently of the impolicy of interfering, without necessity, between these two classes, it may be observed, that the first class, or the makers of machinery, are, as a body, far more intelligent than those who only use it; and though, at present, they are not nearly so numerous, yet, when the removal of the prohibition which cramps their ingenuity shall have had time to operate, there appears good reason to believe, that their number will be greatly increased, and may, in time, even surpass that of those who use machinery. /

438. The advocates of these prohibitions in England seem to rely greatly upon the possibility of preventing the knowledge of new contrivances from being conveyed to other countries; and they take much too limited a view of the possible, and even probable, improvements in mechanics.

439. For the purpose of examining this question, let us consider the case of two manufacturers of the same article, one situated in a country in which labour is very cheap, the machinery bad, and the modes of transport slow and expensive; the other engaged in manufacturing in a country in which the price of labour is very high, the machinery excellent, and the means of transport expeditious and economical. Let them both send their produce to the same market,

and let each receive such a price as shall give to him the profit ordinarily produced by capital in his own country. It is almost certain that in such circumstances the first improvement in machinery will occur in the country which is most advanced in civilization; because, even admitting that the ingenuity to contrive were the same in the two countries, the means of execution are very different. The effect of improved machinery in the rich country will be perceived in the common market, by a small fall in the price of the manufactured article. This will be the first intimation to the manufacturer of the poor country, who will endeavour to meet the diminution in the selling price of his article by increased industry and economy in his factory; but he will soon find that this remedy is temporary, and that the market-price continues to fall. He will thus be induced to examine the rival fabric, in order to detect, from its / structure, any improved mode of making it. If, as would most usually happen, he should be unsuccessful in this attempt, he must endeavour to contrive improvements in his own machinery, or to acquire information respecting those which have been made in the factories of the richer country. Perhaps after an ineffectual attempt to obtain by letters the information he requires, he sets out to visit in person the factories of his competitors. To a foreigner and rival manufacturer such establishments are not easily accessible; and the more recent the improvements, the less likely he will be to gain access to them. His next step, therefore, will be to obtain the knowledge he is in search of from the workmen employed in using or making the machines. Without *drawings*, or an examination of the *machines* themselves, this process will be slow and tedious; and he will be liable, after all, to be deceived by artful and designing workmen, and be exposed to many chances of failure. But suppose he returns to his own country with perfect drawings and instructions, he must then begin to construct his improved machines: and these he cannot execute either so cheaply or so well as his rivals in the richer countries. But after the lapse of some time, we shall suppose the machines thus laboriously improved, to be at last completed, and in working order.

440. Let us now consider what will have occurred to the manufacturer in the rich country. He will, in the first instance, have realized a profit by supplying the home market, at the usual price, with an article which it costs him less to produce; he will then reduce the price both in the home and foreign market, / in order to produce a more extended sale. It is in this stage that the manufacturer in the poor

country first feels the effect of the competition; and if we suppose only two or three years to elapse between the first application of the new improvement in the rich country, and the commencement of its employment in the poor country, yet will the manufacturer who contrived the improvement (even supposing that during the whole of this time he has made only one step) have realized so large a portion of the outlay which it required, that he can afford to make a much greater reduction in the price of his produce, and thus to render the gains of his rivals quite inferior to his own.

441. It is contended that by admitting the exportation of machinery, foreign manufacturers will be supplied with machines equal to our own. The first answer which presents itself to this argument is supplied by almost the whole of the present volume; *That in order to succeed in a manufacture, it is necessary not merely to possess good machinery, but that the domestic economy of the factory should be most carefully regulated.*

The truth, as well as the importance of this principle, is so well established in the Report of a Committee of the House of Commons 'On the Export of Tools and Machinery', that I shall avail myself of the opinions and evidence there stated, before I offer any observations of my own:

Supposing, indeed, that the same machinery which is used in England could be obtained on the Continent, it is the opinion of some of the most intelligent of the witnesses that a want of arrangement / in foreign manufactories, of division of labour in their work, of skill and perseverance in their workmen, and of enterprise in the masters, together with the comparatively low estimation in which the master manufacturers are held on the Continent, and with the comparative want of capital, and of many other advantageous circumstances detailed in the evidence, would prevent foreigners from interfering in any great degree by competition with our principal manufacturers; on which subject the Committee submit the following evidence as worthy the attention of the House:

I would ask whether, upon the whole, you consider any danger likely to arise to our manufactures from competition, even if the French were supplied with machinery equally good and cheap as our own? They will always be behind us until their general habits approximate to ours; and they must be behind us for many reasons that I have before given.

Why must they be behind us? One other reason is, that a cotton manufacturer who left Manchester seven years ago, would be driven out of the market by the men who are now living in it, provided his knowledge had not kept pace with those who have been during that time constantly profiting by the progressive improvements that have taken place in that period; this progressive knowledge and experience is our great power and advantage.

It should also be observed, that the constant, nay, almost daily, improvements which take place in our machinery itself, as well as in the mode of its application, require that all those means and advantages alluded to above, should be in constant operation; and that, in the opinion of several of / the witnesses, although Europe were possessed of every tool now used in the United Kingdom, along with the assistance of English artisans, which she may have in any number, yet, from the natural and acquired advantages possessed by this country, the manufacturers of the United Kingdom would for ages continue to retain the superiority they now enjoy. It is indeed the opinion of many, that if the exportation of machinery were permitted, the exportation would often consist of those tools and machines, which, although already superseded by new inventions, still continue to be employed, from want of opportunity to get rid of them; to the detriment, in many instances, of the trade and manufactures of the country: and it is matter worthy of consideration, and fully borne out by the evidence, that by such increased foreign demand for machinery, the ingenuity and skill of our workmen would have greater scope; and that, important as the improvements in machinery have lately been, they might, under such circumstances, be fairly expected to increase to a degree beyond all precedent.

The many important facilities for the construction of machines and the manufacturing of commodities which we possess, are enjoyed by no other country; nor is it likely that any country can enjoy them to an equal extent for an indefinite period. *It is admitted by everyone, that our skill is unrivalled; the industry and power of our people unequalled; their ingenuity, as displayed in the continual improvement in machinery, and production of commodities, without parallel; and apparently, without / limit.* The freedom which, under our government, every man has, to use his capital, his labour, and his talents, in the manner most conducive to his interests, is an inestimable advantage; canals are cut, and railroads constructed, by the voluntary association of persons whose local knowledge enables them to place them in the most desirable situations; and these great advantages cannot exist under less free governments. These circumstances, when taken together, give such a decided superiority to our people, that no injurious rivalry, either in the construction of machinery or the manufacture of commodities, can reasonably be anticipated.

442. But, even if it were desirable to prevent the exportation of a certain class of machinery, it is abdundantly evident, that, whilst the exportation of other classes is allowed, it is impossible to prevent the forbidden one from being smuggled out; and that, in point of fact, the additional risk has been well calculated by the smuggler.

443. It would appear, also, from various circumstances, that the immediate exportation of improved machinery is not quite so certain as has been assumed; and that the powerful principle of self-interest will urge the makers of it, rather to push the sale in a different direction. When a great maker of machinery has contrived a new machine for any particular process, or has made some great improvement upon those in common use, to whom will he naturally apply for the purpose of selling his new machines? Undoubtedly, in by far the majority of cases, to his nearest and best customers, those to whom he has immediate and / personal access, and whose capability to fulfil any contract is best known to him. With these, he will communicate and offer to take their orders for the new machine; nor will he think of writing to foreign customers, so long as he finds the home demand sufficient to employ the whole force of his establishment. Thus, therefore, the machine-maker is himself interested in giving the first advantage of any new improvement to his own countrymen.

444. In point of fact, the machine-makers in London greatly prefer home orders, and do usually charge an additional price to their foreign customers. Even the measure of this preference may be found in the evidence before the Committee on the Export of Machinery. It is differently estimated by various engineers; but appears to vary from five up to twenty-five per cent on the amount of the order. The reasons are: 1. If the machinery be complicated, one of the best workmen, well accustomed to the mode of work in the factory, must be sent out to put it up; and there is always a considerable chance of his having offers that will induce him to remain abroad. 2. If the work be of a more simple kind, and can be put up without the help of an English workman, yet for the credit of the house which supplies it, and to prevent the accidents likely to occur from the want of sufficient instruction in those who use it, the parts are frequently made stronger, and examined more attentively, than they would be for an English purchaser. Any defect or accident also would be attended with more expense to repair, if it occurred abroad, than in England.

445. The class of workmen who *make* machinery, / possess much more skill, and are paid much more highly than that class who merely *use it*; and, if a free exportation were allowed, the more valuable class would, undoubtedly, be greatly increased; for, notwithstanding the high rate of wages, there is no country in which it can at this moment be made, either so well or so cheaply as in England. We might,

therefore, supply the whole world with machinery, at an evident advantage, both to ourselves and our customers. In Manchester, and the surrounding district, many thousand men are wholly occupied in making the machinery, which gives employment to many hundred thousands who use it; but the period is not very remote, when the whole number of those who *used* machines, was not greater than the number of those who at present *manufacture* them. Hence, then, if England should ever become a great exporter of machinery, she would necessarily contain a large class of workmen, to whom skill would be indispensable, and, consequently, to whom high wages would be paid; and although her manufacturers might probably be comparatively fewer in number, yet they would undoubtedly have the advantage of being the first to derive profit from improvement. Under such circumstances, any diminution in the demand for machinery, would, in the first instance, be felt by a class much better able to meet it, than that which now suffers upon every check in the consumption of manufactured goods; and the resulting misery would therefore assume a mitigated character.

446. It has been feared, that when other countries have purchased our machines, they will cease to / demand new ones: but the statement which has been given of the usual progress in the improvement of the machinery employed in any manufacture, and of the average time which elapses before it is superseded by such improvements, is a complete reply to this objection. If our customers abroad did not adopt the new machinery contrived by us as soon as they could procure it, then our manufacturers would extend their establishments, and undersell their rivals in their own markets.

447. It may also be urged, that in each kind of machinery a maximum of perfection may be imagined, beyond which it is impossible to advance; and certainly the last advances are usually the smallest when compared with those which precede them: but it should be observed, that these advances are generally made when the number of machines in employment is already large; and when, consequently, their effects on the power of producing are very considerable. But though it should be admitted that any one species of machinery may, after a long period, arrive at a degree of perfection which would render further improvement nearly hopeless, yet it is impossible to suppose that this can be the case with respect to all kinds of mechanism. In fact the limit of improvement is rarely approached, except in extensive branches of national

manufactures; and the number of such branches is, even at present, very small.

448. Another argument in favour of the exportation of machinery, is, that *it would facilitate the transfer of capital to any more advantageous mode of employment which might present itself.* If the exportation of machinery were permitted, there would / doubtless arise a new and increased demand; and, supposing any particular branch of our manufactures to cease to produce the average rate of profit, the loss to the capitalist would be much less, if a market were open for the sale of his machinery to customers more favourably circumstanced for its employment. If, on the other hand, new improvements in machinery should be imagined, the manufacturer would be more readily enabled to carry them into effect, by having the foreign market opened where he could sell his old machines. The fact, that England can, notwithstanding her taxation and her high rate of wages, actually undersell other nations, seems to be well established: and it appears to depend on the superior goodness and cheapness of those raw materials of machinery the metals – on the excellence of the tools – and on the admirable arrangements of the domestic economy of our factories.

449. The different degrees of facility with which capital can be transferred from one mode of employment to another, has an important effect on the rate of profits in different trades and in different countries. Supposing all the other causes which influence the rate of profit at any period, to act equally on capital employed in different occupations, yet the real rates of profit would soon alter, on account of the different degrees of loss incurred by removing the capital from one mode of investment to another, or of any variation in the action of those causes.

450. This principle will appear more clearly by taking an example. Let two capitalists have embarked £10,000 each, in two trades: A in supplying a district with water, by means of a steam-engine / and iron pipes; B in manufacturing bobbin net. The capital of A will be expended in building a house and erecting a steam-engine, which costs, we shall suppose, £3000; and in laying down iron pipes to supply his customers, costing £7000. The greatest part of this latter expense is payment for labour; and if the pipes were to be taken up, the damage arising from that operation would render them of little value, except as old metal; whilst the expense of their removal would be considerable. Let us, therefore, suppose, that if A were obliged to

give up his trade, he could realize only £4000 by the sale of his stock. Let us suppose again that B, by the sale of his bobbin net factory and machinery, could realize £8000 and let the usual profit on the capital employed by each party be the same, say 20 per cent: then we have

	Capital invested	*Money which would arise from sale of machinery*	*Annual rate of profit per cent*	*Income*
	£	£	£	£
Water works	10,000	4000	20	2000
Bobbin net Factory	10,000	8000	20	2000

Now, if, from competition, or any other cause, the rate of profit arising from water-works should fall to 10 per cent, that circumstance would not cause a transfer of capital from the water-works to bobbin net making; because the reduced income from the water-works, £1000 per annum, would still be greater than that produced by investing £4000, (the whole sum arising from the sale of the materials / of the water-works), in a bobbin net factory, which sum, at 20 per cent, would yield only £800 per annum. In fact, the rate of profit, arising from the water-works, must fall to less than 8 per cent before the proprietor could increase his income by removing his capital into the bobbin net trade.

451. In any enquiry into the probability of the injury arising to our manufacturers from the competition of foreign countries, particular regard should be had to the facilities of transport, and to the existence in our own country of a mass of capital in roads, canals, machinery, etc., the greater portion of which may fairly be considered as having repaid the expense of its outlay; and also to the cheap rate at which the abundance of our fuel enables us to produce iron, the basis of almost all machinery. It has been justly remarked by M. de Villefosse, in the memoir before alluded to, that *Ce que l'on nomme en France, la question du prix des fers, est, à proprement parler, la question du prix des bois, et la question, des moyens de communications interieures par les routes, fleuves, rivières et canaux.*[a]

The price of iron in various countries in Europe has been stated in

[a] *Translation:* What in France is called the question of the price of iron, is, properly speaking, the question of the price of wood, and the question of methods of national communication by road, river, and canal.

section 215 of the present volume; and it appears, that in England it is produced at the least expense, and in France at the greatest. The length of the roads which cover England and Wales may be estimated roughly at twenty thousand miles of turnpike, and one hundred thousand miles of road not turnpike. The internal water communication of England and France, as far as I have been able to collect information on the subject, may be stated as follows: /

In France

	Miles in length
Navigable rivers	4668
Navigable canals	915·5
Navigable canals in progress of execution (1824)	1388
	6971·5*

But, if we reduce these numbers in the proportion of 3·7 to 1, which is the relative area of France as compared with England and Wales, then we shall have the following comparison:

		England†	*Portion of France equal in size to England and Wales*
		Miles	*Miles*
Navigable rivers		1275·5	1261·6
Tidal navigation‡		545·9	
Canals, direct	2023·5		
Canals, branch	150·6		
	2174·1	2174·1	247·4
Canals commenced		—	375·1
Total		3995·5	1884·1
Population in 1831		13,894,500	8,608,500 /

* This table is extracted and reduced from one of *Ravinet, Dictionnaire Hydrographique*, 2 vols. 8vo. Paris, 1824.

† I am indebted to F. Page, Esq. of Speen, for that portion of this table which relates to the internal navigation of England. Those only who have themselves collected statistical details can be aware of the expense of time and labour, of which the few lines it contains are the result.

‡ The tidal navigation includes: the Thames, from the mouth of the Medway; the Severn, from the Holmes; the Trent, from Trent Falls in the Humber; the Mersey, from Runcorn Gap.

This comparison, between the internal communications of the two countries, is not offered as complete; nor is it a fair view, to contrast the wealthiest portion of one country with the whole of the other: but it is inserted with the hope of inducing those who possess more extensive information on the subject, to supply the facts on which a better comparison may be instituted. The information to be added, would consist of the number of miles in each country, of seacoast, of public roads, of railroads, of railroads on which locomotive engines are used.[a]

452. One point of view, in which rapid modes of conveyance increase the power of a country, deserves attention. On the Manchester Railroad, for example, above half a million of persons travel annually; and supposing each person to save only one hour in the time of transit, between Manchester and Liverpool, a saving of five hundred thousand hours, or of fifty thousand working days, of ten hours each, is effected. Now this is equivalent to an addition to the actual power of the country of one hundred and sixty-seven men, without increasing the quantity of food consumed; and it should also be remarked, that the time of the class of men thus supplied, is far more valuable than that of mere labourers. /

[a] The following appears in the section of additions, p. [399], in the original:
The length of all the roads in the Prussian monarchy amounts at present (1833) to 6766 English miles. The length of all the navigable rivers, excepting those parts joining the sea, is about 2800 English miles.

CHAPTER XXXV

ON THE FUTURE PROSPECTS OF MANUFACTURES, AS CONNECTED WITH SCIENCE

453. In reviewing the various processes offered as illustrations of those general principles which it has been the main object of the present volume to support and establish, it is impossible not to perceive that the arts and manufactures of the country are intimately connected with the progress of the severer sciences; and that, as we advance in the career of improvement, every step requires, for its success, that this connection should be rendered more intimate.

The applied sciences derive their facts from experiment; but the reasonings, on which their chief utility depends, are the province of what is called abstract science. It has been shown, that the division of labour is no less applicable to mental productions than to those in which material bodies are concerned; and it follows, that the efforts for the improvement of its manufactures which any country can make with the greatest probability of success, must arise from the combined exertions of all those most skilled in the theory, as well as in the practice of the arts; each labouring in that department for which his natural capacity and acquired habits have rendered him most fit.

454. The profit arising from the successful application to practice of theoretical principles, will, in / most cases, amply reward, in a pecuniary sense, those by whom they are first employed; yet even here, what has been stated with respect to *patents*, will prove that there is room for considerable amendment in our legislative enactments: but the discovery of the great principles of nature demands a mind almost exclusively devoted to such investigations; and these, in the present state of science, frequently require costly apparatus, and exact an expense of time quite incompatible with professional avocations. It becomes, therefore, a fit subject for consideration, whether it would not be politic in the State to compensate for some of those privations, to which the cultivators of

the higher departments of science are exposed; and the best mode of effecting this compensation, is a question which interests both the philosopher and the statesman. Such considerations appear to have had their just influence in other countries, where the pursuit of science is regarded as a profession, and where those who are successful in its cultivation are not shut out from almost every object of honourable ambition to which their fellow countrymen may aspire. Having, however, already expressed some opinion upon these subjects in another publication,* I shall here content myself with referring to that work.

455. There was, indeed, in our own country, one single position to which science, when concurring with independent fortune, might aspire, as conferring rank and station, an office deriving, in the estimation of the public, more than half its value from the commanding knowledge of its possessor; and it is / extraordinary, that even that solitary dignity – that barony by tenure in thc world of British science – the chair of the Royal Society, should have been coveted for adventitious rank. It is more extraordinary, that a Prince, distinguished by the liberal views he has invariably taken of public affairs – and eminent for his patronage of every institution calculated to alleviate those miseries from which, by his rank, he is himself exempted – who is stated by his friends to be the warm admirer of knowledge, and most anxious for its advancement, should have been so imperfectly informed by those friends, as to have wrested from the head of science, the only civic wreath which could adorn its brow.†

* *Reflections on the Decline of Science in England, and on some of its Causes.* 8vo. 1830. Fellowes.

† The Duke of Sussex was proposed as President of the Royal Society in opposition to the wish of the Council – in opposition to the public declaration of a body of Fellows, comprising the largest portion of those by whose labours the character of English science had been maintained. The aristocracy of rank and of power, aided by such allies as it can always command, set itself in array against the prouder aristocracy of science. Out of about seven hundred members, only two hundred and thirty balloted; and the Duke of Sussex had a majority of eight. Under such circumstances, it was indeed extraordinary, that His Royal Highness should have condescended to accept the fruits of that doubtful and inauspicious victory.

The circumstances preceding and attending this singular contest have been most ably detailed in a pamphlet <by W. H. Fitton>, entitled *A Statement of the Circumstances connected with the late Election for the Presidency of the Royal Society*, 1831, printed by R. Taylor, Red Lion Court, Fleet Street. The whole tone of the tract is strikingly contrasted with that of the productions of some of those persons by whom it was His Royal Highness's misfortune to be supported.

In the meanwhile the President may learn, through the only medium by which his elevated station admits / approach, that those evils which were anticipated from his election, have not proved to be imaginary, and that the advantages by some expected to result from it, have not yet become apparent. It may be right also to state, that whilst many of the inconveniences, which have been experienced by the President of the Royal Society, have resulted from the conduct of his own supporters, those who were compelled to differ from him, have subsequently offered no vexatious opposition: they wait in patience, convinced that the force of truth must ultimately work its certain, though silent course; not doubting that when His Royal Highness is correctly informed, he will himself be amongst the first to be influenced by its power.

456. But younger institutions have arisen to supply the deficiencies of the old; and very recently a new combination, differing entirely from the older societies, promises to give additional steadiness to the future march of science. The *British Association for the Advancement of Science*, which held its first meeting at York* in the year 1831, would have acted as a powerful ally, even if the Royal Society were all that it might be: but in the present state of that body such an association is almost necessary for the purposes of science. The periodical assemblage of persons, pursuing the same or different branches of knowledge, always produces an excitement which is favourable to the development / of new ideas; whilst the long period of repose which succeeds, is advantageous for the prosecution of the reasonings or the experiments then suggested; and the recurrence of the meeting in the succeeding year, will stimulate the activity of the enquirer, by the hope of being then enabled to produce the successful result of his labours. Another advantage is, that such meetings bring together a much larger number of persons actively engaged in science, or placed in positions in which they can contribute to it, than can ever be found at the ordinary meetings of other institutions, even in the most populous capitals; and combined effort towards any particular object can thus be more easily arranged.

457. But perhaps the greatest benefit which will accrue from these assemblies, is the intercourse which they cannot fail to promote between the different classes of society. The man of science will

* The second meeting took place at Oxford in June, 1832, and surpassed even the sanguine anticipations of its friends. The third annual meeting will take place at Cambridge in June, 1833.

derive practical information from the great manufacturers – the chemist will be indebted to the same source for substances which exist in such minute quantity, as only to become visible in most extensive operations – and persons of wealth and property, resident in each neighbourhood visited by these migratory assemblies, will derive greater advantages than either of those classes, from the real instruction they may procure respecting the produce and manufactures of their country, and the enlightened gratification which is ever attendant on the acquisition of knowledge.* /

458. Thus it may be hoped that public opinion shall be brought to bear upon the world of science; and that by this intercourse light will be thrown upon the characters of men, and the pretender and the charlatan be driven into merited obscurity. Without the action of public opinion, any administration, however anxious to countenance the pursuits of science, and however ready to reward, by wealth or honours, those whom they might think most eminent, would run the risk of acting like the blind man recently couched, who, having no mode of estimating degrees of distance, mistook the nearest and most insignificant for the largest objects in nature: it becomes, therefore, doubly important, that the man of science should mix with the world.

459. It is highly probable that in the next generation, the race of scientific men in England will spring from a class of persons altogether different from that which has hitherto scantily supplied them. Requiring, for the success of their pursuits, previous education, leisure, and fortune, few are so likely to unite these essentials as the sons of our wealthy manufacturers, who, having been enriched by their own exertions, in a field connected with science, will be ambitious of having their children distinguished in its ranks. It must, however, be admitted, that this desire in the parents would acquire great additional intensity, if worldly honours occasionally followed successful efforts; and that the country would thus gain for / science, talents which are frequently rendered useless by the unsuitable situations in which they are placed.

460. The discoverers of iodine and bromine, two substances hitherto undecompounded, were both amongst the class of manufacturers,

* The advantages likely to arise from such an association, have been so clearly stated in the address delivered by the Rev. Mr Vernon Harcourt, at its first meeting, that I would strongly recommend its perusal by all those who feel interested in the success of English science. Vide *First Report of the British Association for the Advancement of Science*. York, 1832.

one being a maker of saltpetre at Paris, the other a manufacturing chemist at Marseilles; and the inventor of balloons filled with rarefied air, was a paper manufacturer near Lyons. The descendants of Mongolfier, the first aerial traveller, still carry on the establishment of their progenitor, and combine great scientific knowledge with skill in various departments of the arts, to which the different branches of the family have applied themselves.

461. Chemical science may, in many instances, be of great importance to the manufacturer, as well as to the merchant. The quantity of Peruvian bark which is imported into Europe is very considerable; but chemistry has recently proved that a large portion of the bark itself is useless. The alkali Quinia which has been extracted from it, possesses all the properties for which the bark is valuable, and only forty ounces of this substance, when in combination with sulphuric acid, can be extracted from a hundred pounds of the bark. In this instance then, with every ton of useful matter, thirty-nine tons of rubbish are transported across the Atlantic.

The greatest part of the sulphate of quinia now used in this country is imported from France, where the low price of the alcohol, by which it is extracted from the bark, renders the process cheap; but it cannot be doubted, that when more settled forms of government shall have given security to / capital, and when advancing civilization shall have spread itself over the states of Southern America, the alkaline medicine will be extracted from the woody matter by which its efficacy is impaired, and that it will be exported in its most condensed form.

462. The aid of chemistry, in extracting and in concentrating substances used for human food, is of great use in distant voyages, where the space occupied by the stores must be economized with the greatest care. Thus the essential oils supply the voyager with flavour; the concentrated and crystallized vegetable acids preserve his health; and alcohol, when sufficiently diluted, supplies the spirit necessary for his daily consumption.

463. When we reflect on the very small number of species of plants, compared with the multitude that are known to exist, which have hitherto been cultivated, and rendered useful to man; and when we apply the same observation to the animal world, and even to the mineral kingdom, the field that natural science opens to our view seems to be indeed unlimited. These productions of nature, varied and innumerable as they are, may each, in some future day, become

the basis of extensive manufactures, and give life, employment, and wealth, to millions of human beings. But the crude treasures perpetually exposed before our eyes, contain within them other and more valuable principles. All these, likewise, in their numberless combinations, which ages of labour and research can never exhaust, may be destined to furnish, in perpetual succession, new sources of our wealth and of our happiness. Science and knowledge are subject, in their extension and increase, to laws quite opposite to / those which regulate the material world. Unlike the forces of molecular attraction, which cease at sensible distances; or that of gravity, which decreases rapidly with the increasing distance from the point of its origin; the further we advance from the origin of our knowledge, the larger it becomes, and the greater power it bestows upon its cultivators, to add new fields to its dominions. Yet, does this continually and rapidly increasing power, instead of giving us any reason to anticipate the exhaustion of so fertile a field, place us at each advance, on some higher eminence, from which the mind contemplates the past, and feels irresistibly convinced, that the whole, already gained, bears a constantly diminishing ratio to that which is contained within the still more rapidly expanding horizon of our knowledge.

464. But, if the knowledge of the chemical and physical properties of the bodies which surround us, as well as our imperfect acquaintance with the less tangible elements, light, electricity, and heat, which mysteriously modify or change their combinations, concur to convince us of the same fact; we must remember that another and a higher science, itself still more boundless, is also advancing with a giant's stride, and having grasped the mightier masses of the universe, and reduced their wanderings to laws, has given to us in its own condensed language, expressions, which are to the past as history, to the future as prophecy. It is the same science which is now preparing its fetters for the minutest atoms that nature has created: already it has nearly chained the ethereal fluid, and bound in one harmonious system all the intricate and splendid phenomena of light. It is the science of / *calculation* – which becomes continually more necessary at each step of our progress, and which must ultimately govern the whole of the applications of science to the arts of life.

465. But perhaps a doubt may arise in the mind, whilst contemplating the continually increasing field of human knowledge, that the weak arm of man may want the physical force required to render that

knowledge available. The experience of the past, has stamped with the indelible character of truth, the maxim, that *knowledge is power.* It not merely gives to its votaries control over the mental faculties of their species, but is itself the generator of physical force. The discovery of the expansive power of steam, its condensation, and the doctrine of latent heat, has already added to the population of this small island, millions of hands. But the source of this power is not without limit, and the coal-mines of the world may ultimately be exhausted. Without adverting to the theory, that new deposits of that mineral are not accumulating under the sea, at the estuaries of some of our larger rivers; without anticipating the application of other fluids requiring a less supply of caloric than water – we may remark that the sea itself offers a perennial source of power hitherto almost unapplied. The tides, twice in each day, raise a vast mass of water, which might be made available for driving machinery. But supposing heat still to remain necessary, when the exhausted state of our coal fields renders it expensive: long before that period arrives, other methods will probably have been invented for producing it. In some districts, there are springs of hot water, which have flowed for / centuries unchanged in temperature. In many parts of the island of Ischia, by deepening the sources of the hot springs only a few feet, the water boils; and there can be little doubt that, by boring a short distance, steam of high pressure would issue from the orifice.*

In Iceland, the sources of heat are still more plentiful; and their proximity to large masses of ice, seems almost to point out the future destiny of that island. The ice of its glaciers may enable its inhabitants to liquefy the gases with the least expenditure of mechanical force; and the heat of its volcanoes may supply the power necessary for their condensation. Thus, in a future age, *power* may become the staple commodity of the Icelanders, and of the inhabitants of other volcanic districts;† and possibly the very process by which they will procure this article of exchange for the luxuries of happier climates may, in some measure, tame the tremendous element which occasionally devastates their provinces.

* In 1828, the author of these pages visited Ischia, with a committee of the Royal Academy of Naples, deputed to examine the temperature and chemical constitution of the springs in that island. During the few first days, several springs which had been represented in the instructions as under the boiling temperature, were found, on deepening the excavations, to rise to the boiling point.

† See section 351.

466. Perhaps to the sober eye of inductive philosophy, these anticipations of the future may appear too faintly connected with the history of the past. When time shall have revealed the future progress of our race, those laws which are now obscurely indicated, will then become distinctly apparent; and it may / possibly be found that the dominion of mind over the material world advances with an ever-accelerating force.

Even now, the imprisoned winds which the earliest poet made the Grecian warrior bear for the protection of his fragile bark; or those which, in more modern times, the Lapland wizards sold to the deluded sailors – these, the unreal creations of fancy or of fraud, called at the command of science, from their shadowy existence, obey a holier spell: and the unruly masters of the poet and the seer become the obedient slaves of civilized man.

Nor have the wild imaginings of the satirist been quite unrivalled by the realities of after years: as if in mockery of the College of Laputa, light almost solar has been extracted from the refuse of fish; fire has been sifted by the lamp of Davy; and machinery has been taught arithmetic instead of poetry.

467. In whatever light we examine the triumphs and achievements of our species over the creation submitted to its power, we explore new sources of wonder. But if science has called into real existence the visions of the poet – if the accumulating knowledge of ages has blunted the sharpest and distanced the loftiest of the shafts of the satirist, the philosopher has conferred on the moralist an obligation of surpassing weight. In unveiling to him the living miracles which teem in rich exuberance around the minutest atom, as well as throughout the largest masses of ever-active matter, he has placed before him resistless evidence of immeasurable design. Surrounded by every form of animate and inanimate / existence, the sun of science has yet penetrated but through the outer fold of nature's majestic robe; but if the philosopher were required to separate, from amongst those countless evidences of creative power, one being, the masterpiece of its skill; and from that being to select one gift, the choicest of all the attributes of life; turning within his own breast, and conscious of those powers which have subjugated to his race the external world, and of those higher powers by which he has subjugated to himself that creative faculty which aids his faltering conceptions of a deity, the humble worshipper at the altar of truth would pronounce that being, man; that endowment, human reason.

But however large the interval that separates the lowest from the highest of those sentient beings which inhabit our planet, all the results of observation, enlightened by all the reasonings of the philosopher, combine to render it probable that, in the vast extent of creation, the proudest attribute of our race is but, perchance, the lowest step in the gradation of intellectual existence. For, since every portion of our own material globe, and every animated being it supports, afford, on more scrutinizing enquiry, more perfect evidence of design, it would indeed be most unphilosophical to believe that those sister spheres, obedient to the same law, and glowing with light and heat radiant from the same central source – and that the members of those kindred systems, almost lost in the remoteness of space, and perceptible only from the countless multitude of their congregated globes – should each be no more than a floating chaos of unformed matter; or, being all the work of / the same Almighty Architect, that no living eye should be gladdened by their forms of beauty, that no intellectual being should expand its faculties in decyphering their laws. /

INDEX[a]

[a] Numbers refer to sections; 'c' is Babbage's abbreviation for 'chapter'.

For Product Safety Concerns and Information please contact our EU representative GPSR@taylorandfrancis.com Taylor & Francis Verlag GmbH, Kaufingerstraße 24, 80331 München, Germany

Batch number: 10725800

Printed by Printforce, the Netherlands